# JOURNAL OF CONASENSE

## Communication
## Navigation
## Sensing
## Services

Volume 1, No. 1 (January 2014)

# JOURNAL OF CONASENSE

## Aim

The overall aim of the CONASENSE Journal is to provide a common platform for exchanging ideas among the communities, both the academic and industrial, involved in the fields of Communications, Navigation and Sensing, with emphasis on multidisciplinary views and Smart/Intelligent services that require the effective and efficient integration of these three fields of research and development.

## Scope

The Journal will publish articles on novel research and the latest advances, in the field of communication (in particular, wireless communication), navigation and sensing with special emphasis on the challenges, new concepts and future enablers for the interaction/integration of these technologies for the successful provision of i smart/intelligent services.

The fields of interest include:

- All communications/sensing/navigation systems and techniques, protocols which enable awareness of the physical environment, effective and fast feedback loops between actuation and sensing, a flexible and cognitive architecture which comply with essential requirements like safety, security, near-zero power consumption as well as size, usability and adaptability constraints.
- Control theory aspects in presence of wireless or lossy feedback links (i.e. network control theory), distributed control systems;
- Services and applications such as smart grid, Ambient Assisted Living, Ambient-Intelligence, Smart Cities, Smart Environment, Context-aware services, location-based services, e-Health, but more in general innovative services and applications for contributing to solving societal challenges.
- Data management such as data mining, data retrieval, decision-making algorithms.

*Published, sold and distributed by*:
River Publishers
Niels Jernes Vej 10
9220 Aalborg Ø
Denmark

Tel.: +45369953197
www.riverpublishers.com

*Journal of CONASENSE (Communications, Navigation, Sensing and Services)*
is published three times a year.
Publication programme, 2014: Volume 1 (3 issues)

ISSN:2246-2120 (Print Version)
ISSN:2246-2139 (Online Version)
ISBN:978-87-93102-69-9 (this issue)

# JOURNAL OF CONASENSE

*Volume 1, No. 1 (January 2014)*

# Editorial Foreword: First Issue of the Journal of Communications, Navigation, Sensing and Services (CONASENSE)

Ernestina Cianca and Albena Mihovska

It is our great pleasure to introduce the First Issue of the Journal of the Society on Communication, Navigation, Sensing and Services (CONASENSE). This is a new scientific society with the vision on Communication, Navigation, Sensing and Services (CNSS), 20 to 50 years from now.

The need to carry out cross-cutting research across the Communication/Navigation and Sensing domains has strongly emerged in the last decades. The so-called Cyber-Physical Systems (CPS), which refers to the next generation of embedded ICT systems that are interconnected and collaborating to provide citizens and businesses with a wide range of innovative intelligent services, are examples of systems where communication/sensing and positioning/navigation capabilities are often interconnected. However, the CONASENSE concept is more general and includes also other systems where the requirement of being embedded is more loose. Currently, the typical approach to the integration of communication, navigation and sensing systems follows a bottom-up strategy where different technologies are tied up to provide a single service. The reason is that, commonly, scientists specialized on one field do not have a deep competence on the other field and, hence, cannot manage the CONASENSE concept as a whole. This approach leads to an inefficient integration of communication, navigation and sensing systems which resembles a patchwork where each single piece is still distinguished from each other. We believe that the best strategy for the success of the CONASENSE concept is to follow a top-down approach where new CONASENSE scientists exploit a wider expertise, even if less specialized, which is useful for having a more integrated view of the concept in a first step, and allows to identify problems and develop new solutions in a second step.

The CONASENSE Journal aims to provide a common platform for exchanging ideas among the communities, both the academic and industrial, involved in the fields of Communications, Navigation and Sensing, with emphasis on multidisciplinary views and Smart/Intelligent services that require the effective and efficient integration of these three fields of research and development.

The main motivation behind the CONASENSE initiative, research topics that should be promoted within the CONASENSE initiative and through the journal of the Society, as well as the technical challenges that need to be faced are explained and outlined in the first paper entitled "CONASENSE: Vision, Motivation and Scope" by Ernestina Cianca, AlbenaMihovska, Mauro De Sanctis, Ramjee Prasad.

Moreoer, for this first issue we have selected four papers. Three of them are mainly focused on some specific feature/requirement of the communication architecture needed to support the CONASENSE concept. A keyenabler for this architecture is the flexible availability of radio spectrum, through which information can be transmitted in relation to the service requested, and which is very much relying on proper and accurate sensing techniques. Cognitive radio relies on sensing for better exploitation of the available spectrum, while high-frequency mm-wave bands used in terrestrial and satellite communications are able to satisfy the ever increasing capacity requirements and represent a solution to the limited availability of radio frequency (RF). One proven approach to effectively sense the spectrum is based on collaborative cognitive radio nodes operation. The authors of the paper entitled "Distributed Beamforming with Nodes Selection for Cognitive Radio Networks," propose to employ Distributed Beamforming (DB) at CR networks in order to form beams towards Distant CR (DCR) users so that the CR network is able to forward the signals to DCR users cooperatively. Via adopting the DB method, the CR networks increases its coverage range without causing harmful interferences to PUs.

CR capabilities can also be exploited in Wireless Sensor Networks (WSN), which are traditionally assumedto employ a fixed spectrum allocation and characterized by the communication and processing resource constrains of low-end sensor nodes. The design approaches in present systems based on a regulated flow of power from an energy storage device are not optimal for energy harvesting nodes. Thus, energy is a valuable resource in communications, navigation, sensing and services (CONASENSE)-related applications. The paper entitled "Wireless Sensor and Communication Nodes with Energy Harvesting" by Mehmet ªafak explores methods and novel design approaches to increase the energy efficiency of wireless sensor nodes, which can be vital for applications such as in-body communication. Multipath propagation, shadowing and scattering due to non-homogeneities in the human body may render in-body communications even more difficult.

The realization of CONASENSE faces challenges such as the ability to process the increasing volume of data transmitted by end users, especially in larger urban areas, and collected from large-scale distributed sensor nodes. The availability of novel unified wireless network architectures and infrastructures is crucial. The paper entitled "Towards an Unified Virtual Mobile Wireless Architecture" by Oleg Asenov and Vladimir Poulkov proposes a new unified virtual cell architecture (UVCA)focused on services, applications and content, that incorporates the architectures of device-to-device and machine-to-machine types of communications, thus driving the path towards Future Internet.

The fourth paper entitled "ICT-based Remote Agro-Ecological Monitoring System A Case Study in Taiwan" by Cheng-Long Chuang and Joe-Air Jiang addresses one specific application and proposes a novel approach to remote monitoring in agriculture by use of wireless sensor technology converging with cellular wireless technology for enhanced ubiquitous coverage.

We are confident that this First Issue will give you new incentives and research ideas for the unexplored and promising area of CONASENSE.

The Editors-in-Chief:

Ernestina Cianca and Albena Mihovska

# CONASENSE: Vision, Motivation and Scope

Ernestina Cianca[1], Mauro De Sanctis[1], Albena Mihovska[2]
and Ramjee Prasad[2]

[1]CTIF-Italy center, University of Rome Tor Vergata, via del Politecnico 1, 00133
Rome, Italy, cianca@ing.uniroma2.it
[2]Center for TeleInfrastruktur (CTIF), Aalborg University, Aalborg Denmark,
albena@es.aau.dk, prasad@es.aau.dk

Received September 2013; Accepted November 2013
Publication January 2014

## Abstract

CONASENSE stands for Communication, Navigation, Sensing and Services
and is a new scientific society encouraging cross-cutting research among these
four domains. For each domain, the paper shows examples of the interaction
with the other domains, highlighting recent advances, trends, and challenges,
importance to Future Generation Wireless and other new research areas that
arise from taking a top down approach, with the service on the top and the
available technology seen as a whole on the bottom.

**Keywords:** Network convergence, network architectures, navigation
systems, networked control, sensor networks, wireless access.

## 1 Introduction

The Society on Communication, Navigation, Sensing and Services
(CONASENSE) is a new scientific society focusing on the provision of new
services through the integration of Communication, Navigation and Sensing,
20 to 50 years from now [1].

Convergence of technologies, ultra-high capacity, universal coverage and
maximal energy-and cost-efficiency are key characteristics of the Future
Generation Wireless (FGW) system concept. The key enabling technologies
converging into the FGW system concept are communication, navigation,

*Journal of Communication, Navigation, Sensing and Services, Vol. 1,* 1–22.
doi: 10.13052/jconasense2246-2120.111

sensing and services (CNSS). Sensing is the main enabling technology for the efficient use of available spectrum. Interoperability of numerous heterogeneous devices, their mobility, and capabilities to provide ubiquitous and secure ultra-fast connectivity without harmful interference relies on novel communication technologies and their convergence with navigation and sensing platforms. With the released large amounts of data information coming from large-scale sensor deployments, as well as with the changing user role from a consumer to a 'prosumer', also through the adoption of social networks into the private and business life, more and more data gets generated on a daily basis. Service platforms should be built such that users are guaranteed applications with high QoS and privacy protection but should also abstract the heterogeneity of the communication infrastructure.

In this framework, the CONASENSE society aims to provide a common platform for exchanging ideas among the communities, both academic and industrial, involved in the fields of Communications, Navigation and Sensing, with emphasis on multidisciplinary views and Smart/Intelligent services that require the effective integration of these three fields of research and development.

This paper, which opens the first issue of the CONASENSE journal, aims to explain the main motivation behind this initiative and the research topics that should be promoted within the CONASENSE initiative and through the journal of the Society.

The rest of the paper is organized as follows. Section II aims to give a definition of the CONASENSE concept and a description of the CONASENSE functionalities, i.e. Control (Section III), Sensing (Section IV), Communication (Section V) and Navigation (Section VI). Finally, conclusions are drawn in Section VII.

## 2 Com/Nav Sensing Interaction for Intelligent Services/Systems

Among various anthropological definitions of "intelligence", one is the capability to "adapt" in order to survive to dangers and environmental changes. A system with the capability to adapt can work more efficiently with respect to various criteria and can face future technological breakthrough and new user/service requirements. Efficiency is the capability to perform the same task consuming less resources. For instance, an efficient communication system is a system able to transmit at the same bit rate with less bandwidth, power or complexity.

As a matter of fact, this capability to adapt, or in other words, to react to some measured "state", is a key component of any system/service that aims to provide an answer to some of the current societal challenges [2]. In particular, we need intelligent systems/services for:

- User-centric applications (for instance, patient-centric) where there is the need to adapt to the user requirements in a dynamic environment. In this frame, the user is the entity that is monitored and controlled: an object (e.g. a car in intelligent transportation), a human (e.g. elderly people in assisted living), an animal (e.g. in animal tracking), the environment (e.g. forest fires in environmental monitoring), a process (i.e. industrial automation, smart grid), etc.
- Exploit efficiently current system resources providing the service with the highest quality at the lowest cost, especially when current resources seem to be not sufficient to provide the requested service.

Smart systems are often endowed with some kind of feedback control, as shown in Fig. 1. Usually, there is a centralized or distributed entity able to measure/sense some aspects of the environment/state of the process, process the information and possibly influence the process either directly (with some kind of direct control/actuation) or indirectly by influencing the environment.

Therefore, the provision of smart services and the design of intelligent systems foresee the interaction of communication, navigation and sensing.

Fig. 2 shows the CONASENSE concept of interaction. The provision of enhanced services usually foresees a sensing part, where sensors could be small and distributed in a small or wide area, while in other cases sensors are big Earth Observation satellites. There is also a control part, which contains the intelligence of the system and it is responsible of data fusion and decision making. The nodes of the communication network are sensors, actuators, and the control center besides the various intermediate nodes and gateways of the network. Finally, in all these parts, it is usually crucial to know some "position" or some function of it ("velocity" or "displacement"). For instance, it is important to know the position of sensor nodes to apply energy efficient routing algorithms; it is useful to know the position of communication nodes and eventually optimize their positions according to some criteria; it is for sure important to know the position of the users, for instance, to provide location-based services. Therefore, the position and navigation part (meaning, the part responsible for having awareness on the position of something) is an input to all the other mentioned components.

Process/system

Intelligent agent (it can be distributed): entity able to measure/sense some aspects of the environment/state of the process, process information and possibly influence the environment/process.

**Figure 1**   Scheme of a smart system with feedback control.

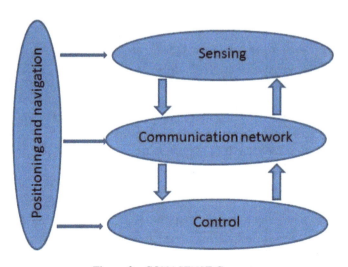

**Figure 2**   CONASENSE Concept

The so-called Cyber-Physical Systems (CPS) [3], which refers to the next generation of embedded ICT systems that are interconnected and collaborating to provide citizens and businesses with a wide range of innovative intelligent services, are examples of systems where communication/sensing and positioning/navigation capabilities are often interconnected as shown in Fig. 2. However, the CONASENSE concept is more general and includes also other systems where the requirement of being embedded is more loose such as surveillance and emergency systems, wide area monitoring, automation etc.

Currently, the typical approach to the integration of communication, navigation and sensing systems follows a bottom-up strategy where different technologies are tied up to provide a single service. The reason is that, commonly, scientists specialized on one field do not have a deep competence on the other field and, hence, cannot manage the CONASENSE concept as a whole. This approach leads to an inefficient integration of communication, navigation and sensing systems which resembles a patchwork where each single piece is still distinguished from each other. We believe that the best strategy for the success of the CONASENSE concept is to follow a top-down approach where new CONASENSE scientists exploit a wider expertise, even if less specialized, which is useful for having a more integrated view of the concept in a first step, and allows to identify problems and develop new solutions in a second step.

The awareness that in any intelligent system/process these three components are interconnected and often in a feedback system, can suggest new approaches to the design of the sensing and control strategy (centralized, distributed, etc.), or of the communication network and on the proper use of positioning. In the next Sections, several examples of the interaction among these three domains are presented with the attempt to suggest also new research areas starting from this novel point of view.

# 3 Control Part

In traditional control systems, information from the sensors is assumed to be instantaneously available for the controller and the control commands are assumed to be instantaneously delivered to the actuator. Moreover, no losses are usually assumed during the delivery of sensed information and commands. These assumptions do not longer hold in control systems with wireless links, where sometimes even the nodes (sensing nodes or controller nodes etc.) are mobile.

Networked control theory is the control theory that takes into account the fact that the feedback channel can introduce delay or losses in a Networked Control System (NCS) [4–6]. An example of a NCS applied to the CONASENSE concept is shown in Fig. 3. In this example, the user $U_1$ is monitored through Sensor $S_1$ and $S_2$ which are connected to the controller through Network $N_1$ and $N_3$. The controller sends control messages to actuators $A_1$ and $A_2$ through network $N_2$ and $N_4$. The sensing signals $s_1$ and $s_2$ are digital signals (discrete-time and discrete-valued) generated by the sensors $S_1$ and $S_2$ are received by the controller as a delayed and modified version of the sensor signals $s'_1$ and $s'_2$. The same holds for the control signals $u_1$ and $u_2$ which are received by the actuators through network $N_2$ and $N_4$. In the ideal case, the received signal $s'_1$, $s'_2$, $u'_1$, $u'_2$, are a perfect replica of the signals $s_1, s_2, u_1, u_2$, without delay $t_1 = t_2 = t_3 = t_4 = 0$.

The behavior of a NCS is affected by:

- Sampling rate constraints and resulting distortions of signals from the sensors and actuators (quantization etc.)
- Bandwidth for communications
- Disturbances in the communication links
- Time delay in the measurement and control loops
- Data errors or package drop.

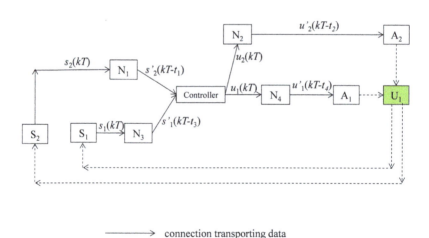

$\longrightarrow$  connection transporting data

$\dashrightarrow$  interaction without transport of data

**Figure 3**  Example of a networked control system.

NC theory is a well established research area where communication and control experts need to interact. Several works can be found in literature, where the two main approaches are [6]:

- To build good communication networks in which the above side effects are trivial or can be neglected.
- To design intelligent controller which can tolerate the above side effects to a certain degree

In both cases, useful theoretical tools can be found by the combination of information theory with control theory [7]. Furthermore, the complex decision making process should be assisted by data mining and data fusion techniques which allows to manage a huge amount of heterogeneous measures from sensors with the objective to extract meaningful information [8].

However, not only the interaction between communication and control is important in the general framework shown in Fig. 2. Also the sensing strategies could be crucial in meeting the control requirements. From the control point of view, it would be better to have a centralized sensing, or a distributed sensing? How does the accuracy in the measurements impact the control?

Here a list of some questions that still need to be answered and that need a multidisciplinary approach:

- How to design MAC protocols that finds an appropriate compromise between satisfying control requirements and use the smallest energy?
- Is it possible to enable the existing controller for NCS usage via a minimum communication support?

  Or minimum sensing support?

- Is it possible to adapt the control strategy to the communication channel? For instance, the delay is too big? Then, that command is useless and another "sub-optimal" command could be sent.
- What is the best control strategy under the system non idealities: uncertainty of measures, errors on data, delay?
- Complex decision making or artificial intelligence algorithms can run over a smartphone?
- Can control algorithms be used also to manage the sensors with the aim to focus the sensing capabilities to the object/situation that is currently critical?

Moreover, most of the work on NCS that can be found in literature is mainly related to automated control in industrial processes. Much less can be found on intelligent systems where there is a human being either at the control side or at the user side: often there is a human operator with a HMI; in some other cases, almost real-time automatic commands must be sent to control or change the behavior of a human being (i.e. Ambient Intelligence). The presence of the human being increases the uncertainty on the effectiveness of the output of the control action as the emotional state of the human being for instance might change his reaction, the reaction time or the type of decision. Therefore, the understanding and modeling of the reaction of the human being in a feedback control system is another important area of research which is receiving more and more attention [9].

## 4 Sensing Part

Sensing can be done by small tiny and low energy sensors, or big sensors (such a ground-based radar or Earth-Observation satellites).

Earth observation data sets are growing in size and variety at an exceptionally fast rate, posing both challenges and opportunities for their access, application and archive. Moreover, wireless sensor networks have changed the paradigm of how sensor data information is made available today. Sensing is not done anymore by few complex sensor devices, wired connected to some control interface, but by a huge number of low-cost, tiny, untethered, battery-powered low-cost MEMS (micro-electro-mechanical systems) devices with limited on-board processing capabilities, storage and short-range wireless communication links based on radio technology, as well as sensing capabilities. Therefore, both data collection and processing are no longer centralized but distributed [10] and the best compromise between local processing and estimation accuracy must be found taking into account the other important metrics such as power consumption and reduction of data transmission. The concept of distributing the "sensing" to different "smaller" sensors, wirelessly connected in an ad hoc manner, is interesting also for space applications. The so-called space-based wireless sensor networks (WSNs) might play an important role in applications such as large-scale space observations, distributed imaging, remote monitoring for deep space exploration, cooperative sensing for high-resolution, synthetic-aperture radar [11].

### 4.0.1 Sensing and control

More generally, large volumes of data are collected by EO satellites, information-sensing mobile devices, aerial sensory technologies and wireless sensor networks. These data must be organized and delivered to make use of it, and eventually to use them for some kind of "control". One of the main current issues is related to the concrete possibility to use it by getting the right and timely information from a huge amount of collected data [12]. Moreover, these data are heterogeneous in nature (images, data, text etc.) or collected by heterogeneous sensors, thus calling for more and more sophisticated and efficient data fusion algorithms. One of the current challenges is related to the possibility to include also information from social networks and more in general, open source data [13].

## 4.0.2 Sensing and positioning

The knowledge on the position of the nodes is essential to optimize the energy efficiency in WSNs, for instance, through proper routing algorithms [14] or topology-based power management strategies [15] or by optimizing the path of a mobile sink that collects data from a large set of sensors distributed over a wide area (for instance, for environmental monitoring applications) [16].

Moreover, there is a subset of WSN in which the geospatial content of the information collected, aggregated, analyzed, and monitored is of fundamental importance [17]. Those are the so-called geosensor networks (GSN) which are usually used to monitor phenomena in geographic space, and there are a key element of any location-based service.

In these WSNs, the spatial aspect might be dominant at different levels:

- Content level, as it may be the dominant content of the information collected by the sensors (e.g. sensors recording the movement or deformation of objects), or
- Analysis level, as the spatial distribution of sensors may provide the integrative layer to support the analysis of the collected information (e.g. analyzing the spatial distribution of chemical leak feeds to determine the extent and source of a contamination).

GSNs enable the Ambient Spatial Intelligence (AmSI), which is concerned with embedding the intelligence to respond to spatiotemporal queries and monitor geographical events in built and natural environments. In [18] it is argued that decentralized spatial computing, where spatial information is partially or completely filtered, summarized, or analysed in a geosensor network, is a fundamental technique required to support AmSI, in contrast with the centralized approach to computation, where global spatial data is collated and processed, for example in a spatial database or GIS.

### 4.0.3 Emerging trends

Some of the key requirements of well-designed sensing networks are: minimized number of sensing elements and measurements, reduced complexity of the sensing method, optimized use of the power. To achieve these objectives, recently the use of compressed sensing has been proposed. Compressed sensing is an emerging theory that is based on the fact that a signal can be recovered through a relatively small number of random projections which contain most of its salient information [19].

In this framework, a new element that is emerging is that the user becomes the sensor and the actuator. For instance, the same "personal" device (i.e. smartphone) can be used as sensor (smartphones already includes many types of sensors) and actuator (alarm, vibration). The innovative concept of people as sensors defines a measurement model, in which measurements are not only taken by calibrated hardware sensors, but in which also humans can contribute their individual 'measurements' such as their subjective sensations, current perceptions or personal observations [20]. These are shared for instance, through social networks. This novel concept could greatly contribute to overcome the challenge to analyze our surroundings in real-time due to the sparsely available data sources. However, it also poses challenges:

- Sensors (human being) are highly mobile.
- What is the density of people that I need to obtain a certain accuracy?
- What is the balance between sensing accuracy and resource usage (number of people involved, network bandwidth, battery usage)
- What are the incentives for people to collaboratively partecipate to the sensing?

## 5 Communication Part

The communication architecture has an important responsibility in fulfilling the requirements of most of the CONASENSE systems, such as very low power consumption, flexibility/adaptability, security and privacy, real-time or almost real-time response to dynamic and complex situations while preserving control, system safety and reliability features. In particular, wireless communication will play a crucial role. In many novel applications, it is getting fundamental the capability of "smart objects" to communicate without the human intervention, which is the so-called Machine-to-Machine (M2M) communication paradigm. M2M applications have their own very unique features [21]: group-based communications, low or no mobility, time-controlled, time-tolerant, small data transmission, secure connection, monitoring, priority alarm messages. These service requirements dictate the architectural design of the communication network.

In this scenario, satellite communications has the potential to play an important role for different reasons. First of all, the "interconnected devices" are remote in many applications or they are dispersed over a wide geographical area or they are inaccessible. Secondly, satellite could provide an alternative path when redundant communication in required, such as when very high reliability must be provided. As a matter of fact, M2M communication via satellite is a reality and represents a real great opportunity for the satellite market. Nevertheless, interoperability with in-situ sensors/actuators, interoperability with the terrestrial network still poses several challenges.

To enable the dynamic control of wireless network resources, while optimizing spectrum use and energy efficiency, one of the key elements of the communication CONASENSE architecture is the cognitive and learning capability.

The communication architecture has also the challenging task to disseminate in an efficient way (efficient with respect to the energy/bandwidth/processing resources) huge amount of data generated by smart sensing systems, supervisory control and data acquisition systems, wide area monitoring systems, and other sensing/monitoring devices throughout large-scale networks. In this framework, gossiping has been proposed [22]. The goal of gossip protocols is to reduce the number of retransmissions by making some of the nodes discard the message instead of forwarding it.

Moreover, the use of data compression techniques is desirable to help mitigating the burden of the communication among sensors and control systems. The information acquired by the sensors should be compressed at the sending terminals as much as possible, before sending through the wireless communication system. The compression should keep the valuable information contained in the data, and the compressed data, when received at receiving terminals, should be perfectly reconstructed for analysis. In [23], the use the Wavelet technology for data compression is proposed.

Two increasingly important requirements that must be fulfilled by the communication architecture are related to security and privacy. On one hand, in many applications (for instance, safety critical applications, monitoring of critical infrastructures etc.), sensing data, information on the position, control commands, must be securely transmitted through several communication networks and protected by any attempt to be jammed, blocked or modified. On the other hand, the user must be able to choose what to share and what to keep private. The privacy issue should be tailored and controlled by the user, since it can change depending on user needs and specific applications/situations. The design of the communication architecture must provide flexible privacy degree controlled by the user.

## 6 Navigation Part

The capability to know the position is crucial for the sensing nodes, for the communication nodes, for the control nodes.

Position and navigation can be done through radio terrestrial networks (i.e. WiFi/cellular) or through Global Navigation Satellite Systems (GNSS).

In previous Sections, several examples have been presented on how positioning information can be used to optimize the design of the sensing system (definition of the topology, distribution of nodes), control strategy and communication protocols (i.e. location-based routing protocols).

On the other hand, in this Section, we outline how sensing, and communication, usually combined in a feedback control, are needed to face the challenges of navigation/positioning systems that arise in many current applications.

For instance, in order to support ITS road safety applications, such as collision avoidance, lane departure warnings and lane keeping, GNSS based position system must provide lane-level (0.5m-1m) or even in lane-level (0.1–03m) lane level accurate and reliable information to users. Current vehicle GNSS receiver (single frequency GPS) can provide road-level accuracy (5–10m). Moreover, in urban areas, the extremely low power of the received GNSS signals and the presence of multipath as a major source of error, makes the GNSS signals not always available and even when available, not reliable.

Regardless those limitations, it would be desirable to use GNSS for those applications thanks to its universal coverage and low equipment cost.

As a matter of fact, accuracy and reliability of GNSS systems can be improved in different ways:

- by combining the signals from GNSS satellites with other information gathered by other types of sensors. It is well known that the combination of sensors such as inertial sensors that measure the motion of the platform to which they are attached without reference to an external system, can improve the availability of the positioning service. By accumulating the measurements from these sensors relative to a known initialization point it is possible to "bridge" the periods when GNSS-derived positions are unavailable, or to improve the confidence in the position estimate when GNSS is available. However, the use of other sensors (baro-altimeters, active RFID tags, cameras) and combination with other information (i.e., maps) has been proposed to improve indoor positioning.

- by combining the signals from GNSS satellites with other RF signals. There is a class of RF signals that are specifically designed for this purpose, and they are the pseudolites and beacons [24]. However, recently is getting more and more attention the use of the so-called signals of opportunity. These are signals not designed for navigation purposes, but that can be used for that purpose. Examples are digital/analog TV, AM radio or nowadays great interest is in the use of WiFi signal [25]. To enable the use of these signals of opportunity, recent advances in wireless communication technologies are crucial. As a matter of fact, a receiver able to use these signals of opportunity must process simultaneously different signals and the capabilities of a software defined cognitive radios to switch quickly frequencies would be important for a practical receiver for positioning purposes.

- broadcasting GNSS corrections generated from a local or regional or global network of ground stations to the user via various data links, mostly 3G networks or communication satellites (i.e. EGNOS). Two example of this approach are represented by the Real Time Kinetic (RTK) or Precise Point Positioning (PPP).

About the latter point, what is the impact of the performance of the communication network used for the broadcast on the GNSS augmented performance? For instance, also GEO satellites such as EGNOS, encounters limitations in urban and rural canyons, accentuated at high latitudes where the EGNOS GEO satellites are seen with low elevation angles. Studies have been done to assess the performance of EGNOS augmented GNSS for road applications [26].

Moreover, what is the impact on the communication network of the added load due to the need to transmit those corrections? This question could be important in future wide-area ITS services. Some studies have been done recently to reduce this load by using proper communication protocols and message format or by assuming less-frequent update of this broadcast information [27].

A communication infrastructure is needed also for broadcasting information for integrity support. Integrity is the measure of trust that can be placed in the correctness of the information supplied by the navigation system. Integrity Support Message (ISM), an integral part of the advanced integrity architectures [28], carries integrity information to the user receiver. The ISM architecture involves:

- Ground monitoring network in charge of collecting the observables to compute the ISM content. We refer to the entity mainly responsible for the computation of the ISM as the Integrity Processing Facility (IPF).
- Broadcast network in charge of delivering the ISM to the final user. The choice of broadcast network will have an impact on the ISM design parameters such as ISM content, ISM update rate, and ISM dissemination latency.

In this context, it is important to assess the impact of the dissemination network performance on the ISM architecture design parameters. By taking into account the performance of the broadcast network, it will be possible to better decide the role of ground station in the provision of integrity monitoring. Moreover, such a study on the communication architecture will also provide guidelines, in terms of ISM contents, ISM packet size, data rate requirements, and ISM update interval. With the objective to provide integrity support via ARAIM to avionic Galileo PRS (Public Regulated Service) users, in [29] a study on the use of TETRA in the ISM dissemination network is performed.

In case of urban road users, multipath is a major source of errors. As another example of research area that calls for a multidisciplinary approach, we wonder if some more local capability to detect the level of multipath, with sensors, cooperation among different users and sensors, could be effectively used to provide either more accuracy or at least a quality measurement of the reliability of the positioning system.

## 7 Conclusion

The need to carry out cross-cutting research across the Communication/Navigation and Sensing domains has strongly emerged in the last decades. Their interaction is the enabler of FGW systems and key component of most of the intelligent services aimed to improve the Quality of Life.

This paper presented an overview of the already existing interactions in terms of trends and challenges, but also provide examples highlighting the importance of taking a top-down approach where the more and more challenging service requirements are met by a joint design of the communication network, sensing, positioning and control strategy.

## References

[1] "Communications, Navigation, Sensing and Services (CONASENSE)", River Publishers, Editor: L.P. Ligthart & R. Prasad, 2012.

[2] http://ec.europa.eu/research/horizon2020

[3] Patricia Derler, Edward A. Lee, Alberto L. Sangiovanni-Vincentelli, "Addressing Modeling Challenges in Cyber-Physical", Technical Report No. UCB/EECS-2011-17, http://www.eecs.berkeley.edu/Pubs/TechRpts/2011/EECS-2011-17.html, March 4, 2011

[4] L. Litz, O. Gabel, I. Solihin, "NCS-Controllers for Ambient Intelligence Networks - Control Performance versus Control Effort" Proc. of the 44$^{th}$ IEEE Conference on Decision and Control, 2005 and 2005 European Control Conference, CDC-ECC '05, Dic. 2005, pp. 1571–1576.

[5] Lixian Zhang, Huijun Gao, Kaynak, O, "Network-Induced Constraints in Networked Control Systems-a Survey" IEEE Transactions on Industrial Informatics, Vol. 9, No. 1, Feb. 2013, pp. 403–416.

[6] Li LI and Fei-Yue WANG, "Control and Communication Synthesis in Networked Control Systems" Int. Journal of Intelligent Control and Systems, Vol. 13, No. 2, June 2008, pp. 81–86.

[7] Massimo Franceschetti and Paolo Minero, "Elements of Information Theory for Networked Control Systems" Information and Control in Networks, Lecture Notes in Control and Information Sciences, Spriger, Vol. 450, 2014, pp 3–37.

[8] Yongmian Zhang, Qiang Ji, Carl G. Looney, "Active Information Fusion For Decision Making Under Uncertainty", Proc. of the Fifth International Conference on Information Fusion, Vol. 1, 8–11 July 2002, pp. 443–450.

[9] Laurel D. Riek and Peter Robinson, "Challenges and Opportunities in Building Socially Intelligent Machines", IEEE Signal Processing Magazine, Vol. 28, No. 3, May 2011, pp. 143–149.

[10] Chongmyung Park, Youngtae Jo, Inbum Jung, "Cooperative Processing Model for Wireless Sensor Network", International Journal of Distributed Sensor Networks, Volume 2013 (2013).

[11] Vladimirova, C.P. Bridges, J.R. Paul, S.A. Malik,. et al. "Space-based wireless sensor networks: Design issues", Proc. of IEEE Aerospace Conf. 2010, March. 2010.

[12] M. Katina, Keith W. Miller, "Big Data: New Opportunities and New Challenges," Computer, June 2013 (Vol. 46, No. 6) pp. 22–24.

[13] R. D. Hull, D. Jenkins, A. McCutchen, "Semantic Enrichment and Fusion of Multi-Intelligent Data,", White Paper Modus Operandi.

[14] M. Tabacchiera, S. Persia, P. Cidronelli, S. Betti, "Routing Optimization for Underwater Optical Networks in Swarm Configuration", Microwave and Optical Technology Letters, 2013.

[15] I. Slama, B. Jouaber, D. Zeghlache, "Topology Contrl and Routing in Large Scale Wireless Sensor Networks", Wireless Sensor Networks, 2010, Vol. 2, pp. 584–598.

[16] ETSI TR 102 300–6, \Terrestrial Trunked Radio (TETRA); Voice plus Data (V+D); Designers' guide; Part 6: Air-Ground-Air," tech. rep., 2011.

[17] A. Stefanidis A., S. Nittel (eds.), GeoSensor Networks, 2004, CRC Press.

[18] M. Duckham, "Decentralized Spatial Computing – Fundation of Geosensor Networks, Springer 2013, ISBN 978-3-642-30852-9, pp. I-XXI, 1–320.

[19] Yiran Shen, Wen Hu, R. Rana, Chun Tung Chou, "Nonuniform Compressive Sensing for Heterogeneous Wireless Sensor Networks,", IEEE Sensors Journal, Vol. 13, No. 6, June, 2013, pp. 2120–2128.

[20] Bernd Resch, "People as Sensors and Collective Sensing- Contextual Observations Complementing Geo-Sensor Network Measurements," J. M. Krisp (ed.), Progress in Location-Based Services, Lecture Notes in Geoinformation and Cartography, DOI: 10.1007/978-3-642-34203-5_22, Springer-Verlag Berlin Heidelberg 2013.

[21] K. Zheng, F. Hu, W. Xiang, M. Dohler, and W. Wang, "Radio Resource Allocation in LTE-Advanced Cellular Networks with M2M Communications," IEEE Communications Magazine, vol. 50, no. 7, pp. 184–192, Jul. 2012.

[22] R. Lanotte, M. Merro, "Semantic Analysis of Gossip Protocols for Wireless Sensor Networks", CONCUR 2011, Lecture Notes on Computer Science, Springer, Vol. 6901, 2011, pp. 156–170.

[23] H. Nikookar, "Wavelet Radio: Adaptive and Reconfigurable Wireless Systems Based on Wavelets," Cambridge University Press, 2013.

[24] J. Barnes, C. Rizos, M. Kanli, D. Small, G. Voigt, N. Gambale, J. Lamance, T. Nunan, C. Reid, "Indoor Industrial Machine Guidance Using Locata: A Pilot Stuy at Bluescope Steel" Proc. ION 2004, Annual Meeting, pp. 533–540, Jun. 2004.

[25] J. Crosby, R. Martin, J.F. Raquet, M. Veth, "Fusion of interial sensors and OFDM Signals of Opportunity for unassisted Navigation," Joint Navigation Conference, June 2010.

[26] M. Obst, R. Schubert, R. Streiter, C. Libeerto, "Benefit Analysis of EGNOS/EDAS for Urban Road Transport Applications," Proc. of the European Conference on ITS, Lyon, 2011.
[27] Ming Qu, "Experimental Studies of Wireless Communication and GNSS Kinematic Positioning Performance in High-Mobility Vehicle Environments," Master Thesis, Faculty of Science and Engineering, Queensland University of Technology, March 2012.
[28] I. Martini, M. Rippl, and M. Meurer, \Integrity Support Message Architecture Design for Advanced Receiver Autonomous Integrity Monitoring," Proceedings of European Navigation Conference, Vienna Austria, 23–25 April 2013.
[29] B. Muhammad, E. Cianca, M. De Sanctis, A. Salonico, F. Rodriguez, "Performance Analysis of Integrity Support Messages Broadcast using TETRA", accepted in ION International Technical Meeting 2014, San Diego, California.

## Biographies

**Ernestina Cianca** received the Laurea degree in Electronic Engineering "cum laude" at the University of L'Aquila in 1997. She got the Ph.D. degree at the University of Rome Tor Vergata in 2001. She concluded her Ph.D. at Aalborg University where she has been employed in the Wireless Networking Groups (WING), as Research engineer (2000–2001) and as Assistant Professor (2001–2003). Since Nov. 2003 she is Assistant Professor in Telecommunications at the URTV (Dpt. of Electronics Engineering), teaching DSP, Information and Coding Theory and Advanced Transmission Techniques. She is the co-director of a II level Master in Advanced Satellite Communication and Navigation Systems. She has been the principal investigator of the WAVE-A2 mission, funded by the Italian Space Agency and aiming to design payloads in W-band for scientific experimental studies of the W-Band channel propagation phenomena and channel quality. She has been coordinator of the scientific activities of the Electronic Engineering Department on the following projects: ESA project European Data Relay System (EDRS); feasibility study for the scientific small mission FLORAD (Micro-satellite FLOwer Constellation of millimeter-wave RADiometers for the Earth and space Observation at regional scale); TRANSPONDER2, funded by ASI, about the design of a payload in Q-band for communications; educational project funded by ASI EduSAT on pico-satellites. She has worked on several European and National projects. Her research mainly concerns wireless access technologies (CDMA and MIMO-OFDM-based systems), integration of terrestrial and satellite systems, short-range communications in biomedical applications. She has been General Chair of the conference ISABEL2010 (Third Symposium on Applied Sciences in Biomedical and Telecommunication Engineering), she has been TPC Co-Chair of the conference European Wireless Technology 2009 (EuWIT2009); TPC Co-Chair in the conference Wireless Vitae 2009. She is Guests Editors of some Special Issues in journals such as Wireless Personal Communications (Wiley) and Journal of Communications (JCM, ISSN 1796–2021). She is author of about 70 papers, on international journals/transactions and proceedings of international conferences.

**Mauro De Sanctis** received the "Laurea" degree in Telecommunications Engineering in 2002 and the Ph.D. degree in Telecommunications and Microelectronics Engineering in 2006 from the University of Roma "Tor Vergata" (Italy).

In autumn of 2004, he joined the CTIF (Center for TeleInFrastruktur), a research center focusing on modern telecommunications technologies located at the University of Aalborg (Denmark).

He was with the Italian Space Agency (ASI) as holder of a two-years research fellowship on the study of Q/V band satellite communication links for a technology demonstration payload, concluded in 2008; during this period he participated to the opening and to the first trials of the ASI Concurrent Engineering Facility (ASI-CEF).

From the end of 2008 he is Assistant Professor at the Department of Electronics Engineering, University of Roma "Tor Vergata" (Italy), teaching "Information and Coding Theory".

From January 2004 to December 2005 he has been involved in the MAG-NET (My personal Adaptive Global NET) European FP6 integrated project and in the SatNEx European network of excellence. From January 2006 to June 2008 he has been involved in the MAGNET Beyond European FP6 integrated project as scientific responsible of WP3/Task3.

In 2006 he was a post-doctoral research fellow for the European Space Agency (ESA) ARIADNA extended study named "The Flower Constellation Set and its Possible Applications".

He has been involved in research activities for several projects funded by the Italian Space Agency (ASI): DAVID satellite mission (DAta and Video Interactive Distribution) during the year 2003; WAVE satellite mission (W-band Analysis and VErification) during the year 2004; FLORAD (Micro-satellite FLOwer Constellation of millimeter-wave RADiometers for the Earth and space Observation at regional scale) during the year 2008; CRUSOE (CRUising in Space with Out-of- body Experiences) during the years 2011/2012.

He has been involved in several Italian Research Programs of Relevant National Interest (PRIN): SALICE (Satellite-Assisted LocalIzation and Communication systems for Emergency services), from October 2008 to September 2010; ICONA (Integration of Communication and Navigation services) from January 2006 to December 2007, SHINES (Satellite and HAP Integrated NEtworks and Services) from January 2003 to December 2004, CABIS (CDMA for Broadband mobile terrestrial-satellite Integrated Systems) from January 2001 to December 2002. In 2007 he has been involved in the Internationalization Program funded by the Italian Ministry of University and Research (MIUR), concerning the academic research collaboration of the Texas A&M University (USA) and the University of Rome "Tor Vergata" (Italy).

He is currently involved in the coordination of scientific activities of the experiments for broadband satellite communications in Q/V band (Alphasat Technology Demonstration Payload 5 - TDP5) funded jointly by ASI and ESA.

He is serving as Associate Editor for the Space Systems area of the IEEE Aerospace and Electronic Systems Magazine. His main areas of interest are: wireless terrestrial and satellite communication networks, satellite constellations (in particular Flower Constellations), resource management of short range wireless systems. He co-authored more than 60 papers published on journals and conference proceedings. He was co-recipient of the best paper award from the 2009 International Conference on Advances in Satellite and Space Communications (SPACOMM 2009).

**Dr Albena Mihovska** obtained the PhD from Aalborg Univcersity, Denmark, where she is currently an Associate Professor at the Center for TeleInfrastruktur (CTIF). She has more than 14 year experience as a researcher in the area of mobile telecommunication systems. She was deeply involved in the design of a next generation radio communication system through her work as the AAU research team leader in the FP6 European funded project WINNER and WINNER II, and later continuing under the CELTIC framework programme as WINNER+, with the related research laying most of the foundations for the current Long-Term Evolution (LTE) and LTE-Advanced, the latter recently approved as an IMT-Advanced standard in ITU-R. She has conducted research activities within the area of advanced radio resource management, cross-layer optimisation, and spectrum aggregation, the results of which were put forward as IMT-A standardisation proposals to the Radio Communication Study Groups of the ITU by the WINNER+ Evaluation Group. She has more than 90 publications including 4 books published by Artech House in 2009 in the next generation mobile communication systems track and 4 book chapters. Further, one of her papers on the topic of next generation communication systems was voted at number 51 of the top 100 IEEE papers for July 2009. She is actively involved in ITU-T Standardization activities within SG13, and Focus Groups Cloud Computing, Smart Grids. She is also actively involved within IEEE Smart Grid Activities. She is a Steering Committee Member of IEEE WCNC, Special Session Chair of GWS2014, and Program Committee Co-Chair for the Wireless Telecommunication Symposium (WTS) 2015. She was Publicity Chair for WPMC2002, Treasurer of IEEE WCNC2006, Secretary of IEEE ComSoc WiMAX 2009 and on the TPC of many highly renowned international conferences, such as IEEE ICC, IEEE VTC, and so forth. She is an Associate Editor of the InderScience International Journal of Mobile Network Design and Innovation (IJMNDI).

**Ramjee Prasad** is currently the Director of the Center for TeleInfrastruktur (CTIF) at Aalborg University, Denmark and Professor, Wireless Information Multimedia Communication Chair.

Ramjee Prasad is the Founding Chairman of the Global ICT Standardisation Forum for India (GISFI: www.gisfi.org) established in 2009. GISFI has the purpose of increasing of the collaboration between European, Indian, Japanese, North-American and other worldwide standardization activities in the area of Information and Communication Technology (ICT) and related application areas. He was the Founding Chairman of the HERMES Partnership – a network of leading independent European research centres established in 1997, of which he is now the Honorary Chair.

He is the founding editor-in-chief of the Springer International Journal on Wireless Personal Communications. He is a member of the editorial board of other renowned international journals including those of River Publishers. Ramjee Prasad is a member of the Steering, Advisory, and Technical Program committees of many renowned annual international conferences including Wireless Personal Multimedia Communications Symposium (WPMC) and Wireless VITAE. He is a Fellow of the Institute of Electrical and Electronic Engineers (IEEE), USA, the Institution of Electronics and Telecommunications Engineers (IETE), India, the Institution of Engineering and Technology (IET), UK, and a member of the Netherlands Electronics and Radio Society (NERG), and the Danish Engineering Society (IDA). He is a Knight ("Ridder") of the Order of Dannebrog (2010), a distinguished award by the Queen of Denmark.

# Distributed Beamforming with Nodes Selection for Cognitive Radio Networks

X.Lian, H. Nikookar and L.P. Ligthart

*Faculty of Electrical Engineering, Mathematics and Computer Science,*
*Microwave Sensing, Systems and Signals (MS3) Group,*
*Delft University of Technology, Mekelweg 4, 2628CD Delft, The Netherlands*
*xiaohualiantud@gmail.com, h.nikookar@tudelft.nl, L.P.Ligthart@tudelft.nl*

Received September 2013; Accepted November 2013
Publication January 2014

## Abstract

In order to collaboratively forward the Cognitive Radio (CR) signal to the Distant CR (DCR) users, we have introduced the Distributed Beamforming (DB) technique CR networks. However we face a practical difficulty of the extreme narrow main beam in the pattern generated by the DB method, when applying it to CR networks. To solve this problem, we propose a novel Nodes Selection (NS) method based on studies of the differences in beam width of a broadside array and an end-fire array. The proposed NS method selects those CR nodes, which are able to form a full size end-fire array and a reduced size broadside array. It chooses CR nodes located in the "belt" area along the direction of the DCR user. Simulation results of the average beampattern of our NS method show that the main beams are successfully directed towards the DCR users and are enlarged for practical applications in CR networks. What is more, for a CR network with a large physical size, our NS method can widen the main beam while maintaining sufficiently low sidelobe levels for CR transmission.

**Keywords:** Cognitive Radio (CR), Distributed Beamforming (DB), Collaborative Beamforming, Cognitive Radio network, Nodes Selection (NS).

*Journal of Communication, Navigation, Sensing and Services, Vol. 1,* 23–46.
doi: 10.13052/jconasense2246-2120.112

# 1 Introduction

Cognitive Radio (CR) is a promising solution to solve the problem of intensive usage of the natural radio resource, spectrum. It has been initially introduced by Joseph Mitola [1], who described how CR could enhance the flexibility of wireless services through a radio knowledge representation language. CR provides various solutions to accommodate this spectrum to be used by unlicensed wireless devices without disrupting the communications of the Primary Users (PUs) of the spectrum [2]. CR capabilities can also be exploited in Wireless Sensor Networks (WSN), which are traditionally assumed to employ a fixed spectrum allocation and characterized by the communication and processing resource constrains of low-end sensor nodes [3]. Depending on the applications, WSN composed of sensor nodes equipped with CR may benefit from its potential advantages, such as dynamic spectrum access and adaptability for reducing power consumptions. A CR network is formed by CR nodes that are geographically distributed in a certain area, which is shown in Figure 1. Those nodes are possibly wireless terminals, subscriber users, or sensors in the CR network. In this paper, we propose to employ Distributed Beamforming at CR networks in order to form beams towards Distant CR users so that the CR network is able to forward the signals to DCR users cooperatively. Via adopting the DB method, the CR networks increase its coverage range without causing harmful interferences to PUs.

DB is also referred to as collaborative beamforming, and is originally employed as an energy-efficient scheme to solve long distance transmission in WSN, in order to reduce the amount of required energy and consequently to extend the utilization time of the sensors [4]. The basic idea of DB is that a set of nodes in the wireless network acts as a virtual antenna array and then forms a beam towards a certain direction to collaboratively transmit a signal. DB has been proposed in [4] and it has been shown that by employing $K$ collaborative nodes, the collaborative beamforming can result in up to $K$-fold gain in the received power at a distant access point. Recently a cross-layer approach for DB in wireless ad-hoc networks has been discussed in [5] applying two communication steps. In the first phase, nodes transmit locally in a random access time-slotted fashion. In the second phase, a set of collaborating nodes, acting as a distributed antenna system, forward the received signal to one or more far away destinations. The improved beam pattern and connectivity properties have been shown in [6], and a reasonable beamforming performance affected by nodes synchronization errors has been discussed in [7]. DB has also been introduced in relay communication systems.

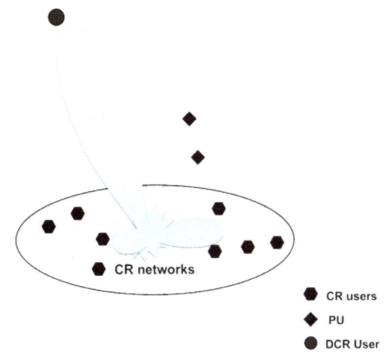

**Figure 1** Coexistence of PUs and CR with DB techniques

Different types of relays have been considered, e.g. Amplify-and-Forward (AF) relays, Filter and Forward (FF) relays, etc. The models discussing relay networks in [8–12] have been proposed to have a source, a relay, and a destination, where transmit DB is employed both at the source and at the relay. The authors in [8–11] have developed several DB techniques for relay networks with flat fading channels. In [12], frequency selective fading has been considered.

DB requires accurate synchronization; in other words, the nodes must start transmitting at the same time, synchronize their carrier frequencies, and control their carrier phases so that their signals can be combined constructively at the destination. A synchronization technique based on time-slotted round-trip carrier synchronization has been proposed for DB in[13], and a review has been given in [14]. In this paper, we adopt the master-slave architecture proposed in [7], where a designated master transmitter (one of CR nodes in the networks) coordinates the synchronization of others (slave) transmitters for DB. This method has also been proved in [7] that a large fraction of DB gains can still be realized even with imperfect synchronization corresponding to phase errors with moderately large variance.

In [4], the authors demonstrate that the sidelobe level in the beampattern generated by the DB method will approach $\frac{1}{K}$. where $K$ is the number of cooperating nodes. Thus as long as PUs are not located in the same direction as the beampattern for the DCR user, the transmit power arriving at PUs' directions will always be $K$-1 times less than that towards the DCR users. As a result, it is not necessary to direct specific nulls towards directions of PUs; the DB method by its nature is capable of guaranteeing $K$-1 times less power towards PUs compared with that towards DCR users. However, to apply the DB technique in CR networks, a practical difficulty rather than nulls directing rises due to the fact that the width of the main beam in the beampattern generated by the DB method is relying on the working frequency of the CR networks. As also shown in [4], the main beam of the beampattern will become narrower when $\widetilde{R}$ increases, where $\widetilde{R} \triangleq R/\lambda$, $R$ is the covering radius of the network and $\lambda$ is the wavelength. If we consider a CR network with $R$=100m and utilizing the spectrum of the Ultra High Frequency (UHF) band, which is, for instance, 750MHz, we can obtain that $\widetilde{R} = 250$. In [4], the authors have shown that the width of the main beam can be approximated by $\frac{35°}{\widetilde{R}}$. Thus in our example, the width of the main beam will become about $0.1°$. Such a narrow main beam implies that once the DOA estimation of the distant nodes is not accurate enough, the main beam in the pattern may miss its direction. Meanwhile, it also reveals that the width of the main beam in the beampattern mostly relies on the center frequency at which the CR network is able to access.

To solve this problem of the extreme narrow main beam, we propose a novel Nodes Selection (NS) method. The presented NS method is based on the differences in beam width of a broadside array and an end-fire array. The beampatterns of these two types of arrays allow us to find out how the main beam of the end-fire array is much wider than that of the broadside array. We thus conclude that the "broadside" size of the CR network should be small so that a beampattern with a wider main beam can be maintained. As a result, we suggest selecting CR nodes, which are able to form a full size end-fire array and a reduced size broadside array. In other words, we choose those CR nodes which are located in the "belt" area along the direction of the distant nodes.

The paper is arranged as follows. Section 2 introduces DB technique to CR networks. A novel NS method to enlarge the main beam of the beampattern generated by the DB technique is presented in section 3. Simulations will be given afterwards in section 4, showing that our NS method is effective in generating a wider main beam in the beam pattern. Section 5 concludes the whole chapter.

## 2 DB for CR Networks

The geometrical structure of the first model together with distant receiver terminals including PUs and DCR users is illustrated in Figure 2. $K$ CR nodes are uniformly distributed over a disc centered at O with radius $R$. We denote the polar coordinates of the $k$th CR node by $(r_k, \Psi_k)$. The number of DCR users is $L_{DCR}$. These DCR users are considered as access points, and located in the same plane at $(A_i^{DCR}, \phi_i^{DCR})$, $i = 1, 2, ..., L_{DCR}$. Meanwhile the number of PUs is $L_{PU}$, and these users coexist with DCR users. Their locations are $(A_i, \phi_i)$, $i = 1, 2, ..., L_{PU}$. The CR nodes in the CR network are requested to form a virtual antenna array and collaboratively transmit a common message $s(t)$.

### 2.1 Necessary assumptions

Without loss of generality, we adopt the following assumptions:

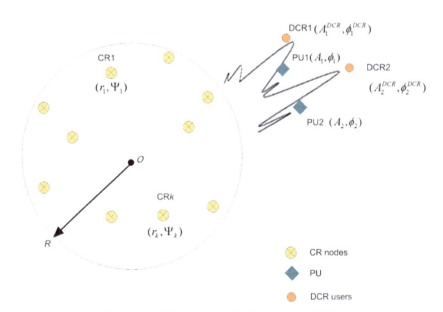

**Figure 2**    CR networks with DCR users and PUs

1) The number of CR nodes are larger than that of DCR users, i.e., $K > L_{DCR}$. This is required to solve the multi beam generating problem which has been discussed in [15].
2) All DCR users and PUs are located in the farfield of the CR network, such that $A_i^{DCR} >> R, i = 1, 2, ..., L_{DCR}$ and $A_i >> R, i = 1, 2, ..., L_{PU}$
3) The bandwidth of $s(t)$ is narrow, so that $s(t)$ is considered to be constant during the time interval $R/c$, where $c$ is the speed of light. However, it has been discussed in [16] that the OFDM scheme is the first recommended candidate for use in the CR network. Since the OFDM signal can be regarded as a combination of narrow band modulated signals, the proposed method in this paper can also be implemented in wide band CR OFDM systems.

## 2.2  DB for CR networks

Let $x_k(t)$ denote the transmitted signal from the $k$th node,

$$x_k(t) = s(t)e^{j2\pi ft} \tag{1}$$

where $f$ is the carrier frequency. The received signal at an arbitrary point $(A, \phi)$ in the far field due to the $k$th node transmission is [17]

$$r_k(t) = \beta_k x_k(t - \frac{d_k}{c}) = \beta_k s(t - \frac{d_k}{c})e^{j2\pi ft}e^{-j\frac{2\pi}{\lambda}d_k} \tag{2}$$

where $d_k$ is the distance between the $k$th node and the access point $(A, \phi)$, and $\beta_k = (d_k)^{\frac{-\gamma}{2}}$ is the signal path loss with $\gamma$ donating the path loss exponent. Making use of assumption 2 in the previous paragraph [17], $\beta_k$ and $d_k$ are approximated by

$$d_k = \sqrt{A^2 + r_k^2 - 2Ar_k\cos(\phi - \Psi_k)} \approx A - r_k\cos(\phi - \Psi_k) \tag{3}$$

$$\beta_k = (d_k)^{\frac{-\gamma}{2}} \approx [A - r_k\cos(\phi - \Psi_k)]^{\frac{-\gamma}{2}} \approx \beta\left(1 + \frac{\gamma r_k\cos(\phi - \Psi_k)}{2A}\right) \tag{4}$$

where $\beta = A^{\frac{-\gamma}{2}}$. It is also ensured that $\frac{\gamma r_k\cos(\phi - \Psi_k)}{2A} << 1$. Thus $\beta_k$ can then be approximated by $\beta$. Substituting equation (3) and (4) into equation (2), it follows that [17],

$$r_k(t) \approx \beta e^{-j\frac{2\pi}{\lambda}A}s(t - \frac{A}{c})e^{j2\pi ft}e^{j\frac{2\pi}{\lambda}r_k\cos(\phi - \Psi_k)} \tag{5}$$

We assume there is only one DCR, i.e., $L_{DCR} = 1$, and simplify $(A_1^{DCR}, \phi_1^{DCR})$ by $(A_0, \phi_0)$. The case with more than one DCR users has been discussed in [15]. As proposed in [4], we adopt for the initial phase of each node

$$\varphi_k = -\frac{2\pi}{\lambda} r_k \cos(\phi_0 - \Psi_k) \tag{6}$$

The received signal $r_k(t)$ at $(A, \phi)$ becomes

$$r_k(t) \approx \beta e^{-j\frac{2\pi}{\lambda}A} s(t - \frac{A}{c}) e^{j2\pi ft} e^{j\frac{2\pi}{\lambda} r_k \cos(\phi - \Psi_k)} e^{-j\frac{2\pi}{\lambda} r_k \cos(\phi_0 - \Psi_k)} \tag{7}$$

The array factor $F(\phi|\, r_k, \Psi_k)$ yields:

$$F(\phi|\, r_k, \Psi_k) \approx \frac{1}{K} \sum_{k=1}^{K} e^{j\frac{2\pi}{\lambda} r_k [\cos(\phi - \Psi_k) - \cos(\phi_0 - \Psi_k)]} \tag{8}$$

$$= \frac{1}{K} \sum_{k=1}^{K} e^{-j\frac{4\pi}{\lambda} r_k \sin(\frac{\phi - \phi_0}{2}) \sin(\frac{\phi + \phi_0 - 2\Psi_k}{2})}$$

We assume there are many CR nodes and the locations of CR nodes follow a uniform distribution over a disk of radius $R$, leading to the probability density functions (pdf)

$$\begin{cases} f_{r_k}(r) = \frac{2r}{R^2}, \ 0 \leq r < R \\ f_{\Psi_k}(\Psi_k) = \frac{1}{2\pi}, \ -\pi \leq \Psi_k < \pi \end{cases} \tag{9}$$

By defining $z_k \triangleq \frac{r_k}{R} \sin\left(\Psi_k - \frac{\phi_1 + \phi_0}{2}\right)$, the compound random variable $z_k$ has a pdf [4]

$$f_{z_k}(z_k) = \frac{2}{\pi} \sqrt{1 - z_k^2}, \ -1 \leq z < 1 \tag{10}$$

The array factor in equation (8) can now be written as

$$F(\phi|z_k) = \frac{1}{K} \sum_{k=1}^{K} \exp\left(-j4\pi \tilde{R} \sin\left(\frac{\phi - \phi_0}{2}\right) z_k\right) \tag{11}$$

where $\tilde{R} \triangleq R/\lambda$ is the radius of the disk normalized by the wavelength. The far field beampattern is defined by

$$P(\phi|z_k) \triangleq |F(\phi|z_k)|^2 \tag{12}$$

and the average array beam pattern of the CR networks becomes [4]

$$P_{av}(\phi) \triangleq \mathrm{E}\left[P(\phi \,|\, z_k)\right] = \frac{1}{K} + \left(1 - \frac{1}{K}\right)\mu^2(\phi) \qquad (13)$$

where

$$\mu(\phi) = E\left[F(\phi)\right] = \left|\frac{2J_1\left(\alpha(\phi)\right)}{\alpha(\phi)}\right| \qquad (14)$$

$$\alpha(\phi) \triangleq 4\pi\widetilde{R}\sin\left(\frac{\phi - \phi_0}{2}\right) \qquad (15)$$

$J_n(\cdot)$ stands for the $n$th order Bessel function of the first kind, and $E[\cdot]$ stands for the statistical expectation. Above equations learn us that if each CR node adopts the initial phase as given in equation (6), the average pattern generated by the whole CR network can be obtained from equation (13).

Figure 3 shows the average beampattern of the DB method for different $K$ ($K = 8, 16$) and $\widetilde{R}(\widetilde{R} = 2, 4, 8)$. We assume that the direction of DCR is $\phi_0 = 0°$. It can be seen that when the beam angle moves away from the

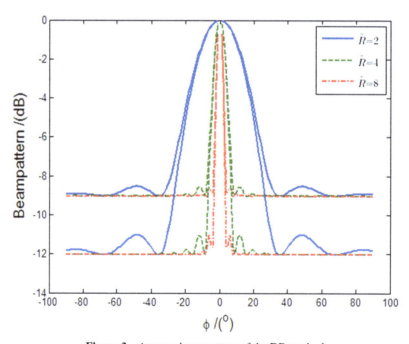

**Figure 3**   Average beampattern of the DB method

direction of DCR, the sidelobe approaches $\frac{1}{K}$, i.e., $10 \log_{10}\left(\frac{1}{8}\right) \approx -9dB$ and $10 \log_{10}\left(\frac{1}{16}\right) \approx -12dB$, respectively. This leads to the logical conclusions that the sidelobe level decreases when $K$ increases, and that the larger $\widetilde{R}$ becomes, the narrower the main beam will be, and consequently the better directivity in the beampattern will be achieved. Figure 3 is only for demonstration purposes, because the value of the parameters that we considered ($K = 8, 16$; $\widetilde{R} = 2, 4, 8$) are not realistic in the real application. In practical applications, $\widetilde{R}$ should be large enough so that CR networks can have enough CR nodes.

## 3 NS for CR Networks with Enlarged Main Beam

In this section we propose a new NS method for CR networks to select proper nodes to achieve wider main beam in order to solve the problem of $\widetilde{R}$ increasing rapidly considering the higher operating frequency of the CR networks. We first consider two extreme cases of the CR network, which are two types of array antennas: broadside array and end-fire array by projecting the location of each CR node along an X and Y axis, and study the properties of these two array antennas. Next we propose the NS method.

We assume one DCR user is located along the X axis ($\phi_0 = 0°$). We then consider the location of a CR node by projecting it into the Cartesian coordinate system (X and Y directions) as shown in Figure 4. In this way we create two virtual arrays: broadside array and end-fire array. We now discuss the performance of the two separate arrays (broadside array and end-fire array) instead of the full CR network.

The average beampatterns of these two arrays are summarized in the following equations

$$\overline{P}_{broadside}(\phi) = \frac{1}{K} + \left(1 - \frac{1}{K}\right)\mu_b^2(\phi) \tag{16}$$

where

$$\mu_b(\phi) = \left|\frac{2J_1(\alpha_b(\phi))}{\alpha_b(\phi)}\right| \tag{17}$$

$$\alpha_b(\phi) = 2\pi\widetilde{R}(\sin\phi - \sin\phi_0) \tag{18}$$

and

$$\overline{P}_{end-fire}(\phi) = \frac{1}{K} + \left(1 - \frac{1}{K}\right)\mu_e^2(\phi) \tag{19}$$

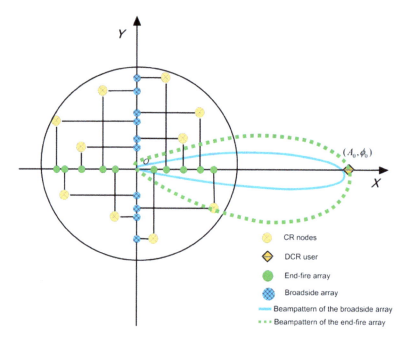

**Figure 4**    Converting locations of CR nodes into broadside and end-fire arrays

where

$$\mu_e(\phi) = \left| \frac{2J_1\left(\alpha_e\left(\phi\right)\right)}{\alpha_e\left(\phi\right)} \right| \tag{20}$$

$$\alpha_e\left(\phi\right) = 2\pi\widetilde{R}\left(\cos\phi - \cos\phi_0\right) \tag{21}$$

Proofs of equations (16) to (21) can be found in Appendix A.

The results of equation (16) and (21) are shown in Figure 5. We assume there are 32 nodes in the CR network and the normalized radius is 35, i.e., $K = 32$ and $\widetilde{R} = 35$. We also assume there is only one DCR user, and its DOA is $\phi_0 = 0°$. We can conclude from Figure 5 that the broadside array has the same average beampattern as the previous CR network. This is due to the result shown in equation (16). The $\alpha_b\left(\phi\right)$ in equation (18) can be approximated to $\alpha(\phi)$ defined in equation (15), when $\phi$ is close to $\phi_0$. As a result in the angle area close to $\phi_0 = 0°$, the performances of the beampattern of the broadside array and the DB method are very similar to each other. In addition, to the same conclusion that has also been drawn in [18], we have discovered that the width of the main beam is in a reverse relationship with the size of the broadside array. Thus we are inspired by these two facts that if

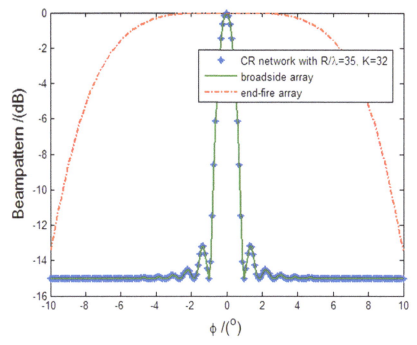

**Figure 5**   Beampattern of CR network, broadside array and end-fire array

we want to enlarge the width of the main beam, we have to decrease the length (size) of the broadside array and we can adopt the end-fire array instead.

Based on this idea, we propose a NS method and select those nodes, which are able to form a full size end-fire array and a reduced size broadside array. Thus we choose the nodes in a relatively narrower belt along the Direction of Arrival (DOA) of the DCR user, as shown in Figure 6. In Figure 6, the CR nodes are selected in a way that the size of the "broadside" is limited to $D$, where $\frac{D}{\lambda} < \widetilde{R}$. When we consider the case with more than one DCR user coexisting with the CR network, e.g. two DCR users, the NS method is demonstrated in Figure 7. We choose those CR nodes which are in the two "belt" areas as shown in Figure 7. In addition, for those double selected CR nodes, which are in the cross area, we adopt the method which let them randomly choose one of the two DCR users to serve, which has been proposed and discussed in [15].

When we consider more than two DCR users, the NS method is slightly different than the case of two DCR users. Figure 8 demonstrates the idea of how to select CR nodes utilizing our proposed NS method, when there are three present DCR users. As we can see from Figure 8, there are three different

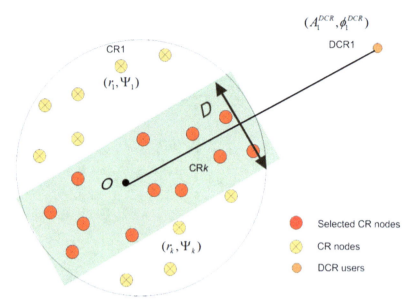

**Figure 6**    NS for CR networks

types of CR nodes in the CR network. They are nodes which are only selected once in the belt area along one DCR user, and are requested to generate main beams towards this DCR user. They are also nodes which are double selected and triple selected. For the double selected nodes, they can choose between two DCR users, and we let them randomly choose towards which DCR user they want to direct their main beam in the pattern. For the triple selected CR nodes, they have three options of different DCR users to choose and again we let them choose among these three. It is worth noticing that double and triple selected nodes have different sets of DCR users to choose from.

## 4 Simulation Results of the NS Method

Figure 9 and 10 demonstrate the selected nodes in the CR networks with different values of $D$ by adopting the proposed NS method. Considering the fact that the number of selected nodes varies from each simulation, we do averaging in order to show a general result of the beampattern. As a result, the beampattern of the NS method shown in Figure 11 is the average beampattern. In Figures 9–10, we assume there is only one DCR user and its DOA is $\phi_0 = 0°$. We choose the nodes within the width of the belt $D = 15\lambda$ and $D = 35\lambda$,

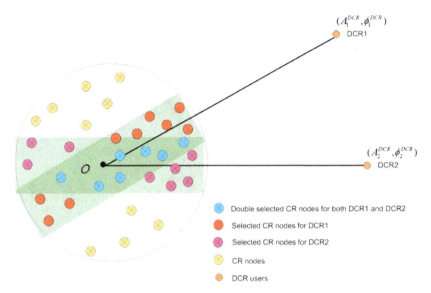

**Figure 7**   NS for CR networks with two DCR users

as shown in Figure 9 and 10, respectively. We assume there are 60 nodes in the network and the normalized radius of the network is 50, i.e., $K = 60$, and $\tilde{R} = 50$. The result of the beampattern shown in Figure 11 is the average of 1000 runs.

Figure 9 and 10 show that those CR nodes, which are located in the belt area, as defined in Figure 6, are successfully selected for transmission. We can also see from these two figures that when $D$ is smaller, less number of CR nodes will be selected for transmission.

Figure 11 shows that after adopting the proposed NS method, the main beam in the beampattern of the DB method is enlarged. Employing the NS method with defined $D = 10\lambda$, $D = 15\lambda$ and $D = 35\lambda$, the main beam is about six times, four times and two times wider than that without adopting the NS method, respectively. However, with smaller $D$, less CR nodes will be selected, as shown in Figure 6. Therefore, the beampattern has a higher sidelobe level than that with a larger $D$, since the asymptotic sidelobe level of the beampattern is proportional to the reverse of the number of nodes of the CR network, as explained in Figure 3. With smaller $D$, which selects fewer CR nodes for transmission, the width of the main beam is the largest. This shows that NS is a trade-off between the width of the main beam and the sidelobe level in the beampattern.

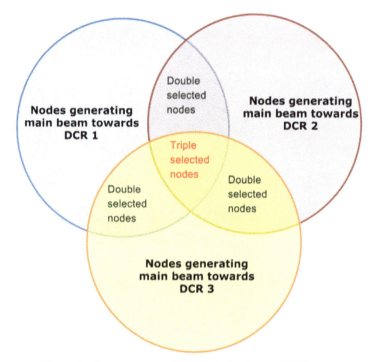

**Figure 8**  Demonstration of NS method with three DCR users

Figure 12 and 13 consider the case with two DCR users, which are located at $\phi_1 = 0°$ and $\phi_2 = 15°$ when $D = 15\lambda$. Figure 12 shows the selected nodes in the CR networks to participate in CR transmission towards two DCR users. We can see in Figure 12 that there are a few CR nodes which are double selected for participating in transmission towards both DCR users.

Figure 13 shows the average beampattern of the NS method for two DCR users. As can be seen from this figure, the two main beams in the beampattern are directed towards $\phi_1 = 0°$ and $\phi_2 = 15°$, respectively. The main beams are both widened via adopting our NS method. When we employ NS method with $D = 15\lambda$, we can see from Figure 13 that both two main beam are broadened to 3°.

Figure 14 and 15 consider the case with three DCR users, which are located at $\phi_1 = 0°$, $\phi_2 = 15°$ and $\phi_3 = -20°$ when $D = 15\lambda$, and show the selected nodes in the CR networks to participate in CR transmission towards these three DCR users. It shows three types of CR nodes as demonstrated in Figure 8. Figure 15 shows the average beampattern of the NS method for three

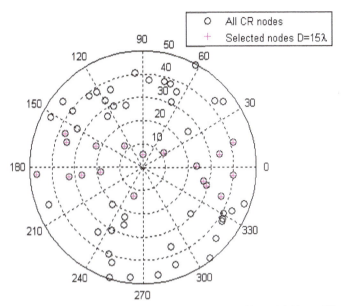

**Figure 9** Selected CR nodes in the CR networks $D = 15\lambda(\phi_0 = 0°)$

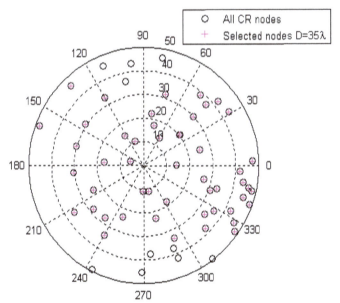

**Figure 10** Selected CR nodes in the CR networks $D = 35\lambda \ (\phi_0 = 0°)$

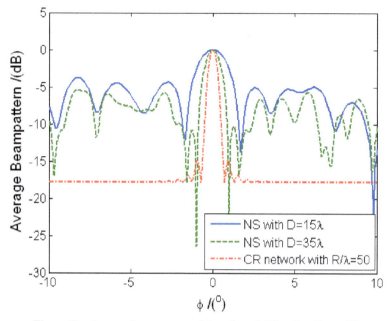

**Figure 11**    Average beampattern of the selected CR nodes ($\phi_0 = 0°$)

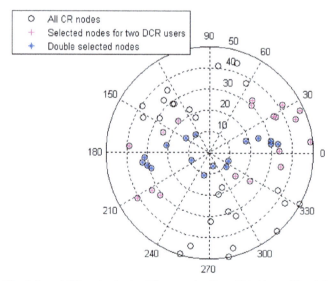

**Figure 12**    Selected CR nodes in the CR networks $D = 15\lambda$ ($\phi_1 = 0°$ and $\phi_2 = 15°$)

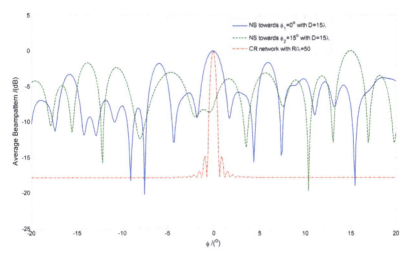

**Figure 13** Average beampattern of the selected CR nodes for two DCR users

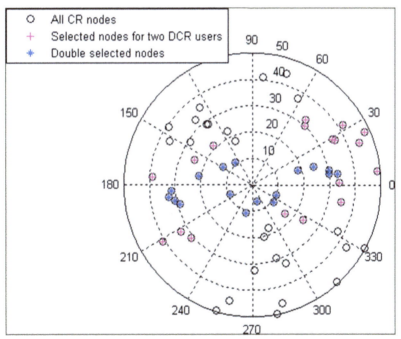

**Figure 14** Selected CR nodes in the CR networks $D = 15\lambda$ ($\phi_1 = 0°$, $\phi_2 = 15°$ and $\phi_3 = -20°$)

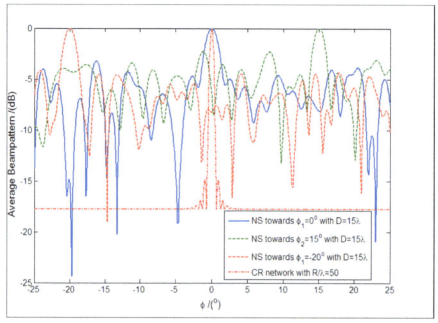

**Figure 15**   Average beampattern of the selected CR nodes for two DCR users

DCR users. It shows that three main beams in the beampattern are directed towards $\phi_1 = 0°$, $\phi_2 = 15°$ and $\phi_3 = -20°$, respectively. All three main beams which are displayed in the pattern towards DCR users are all widened.

Figure 16 shows the result of the average beampattern of our proposed NS method when applied to large CR networks ($\widetilde{R} = 100, 200, 400$). We consider that the distribution density of the CR nodes remains the same with that of the Figure 9–10 ($K = 60$, $\widetilde{R} = 50$), and only the size of the networks are increased. Consequently the number of CR nodes of the considered CR networks in Figure 9 and 10 is $K = \frac{60 \times 100^2}{50^2} = 240$, $K = 960$ and $K = 3840$. We adopt $D = 15\lambda$ for all the three CR networks. It can be seen from this Figure that the main beams are much wider than those of the DB method without NS method which are approximated by $\frac{35°}{R}$. When the size of the CR network is enlarged, more nodes will be selected to participate in CR transmission. As a result sufficiently lower sidelobes can be achieved. For the case of $\widetilde{R} = 400$, far sidelobe levels become approximately $-15dB$, where near sidelobes are higher.

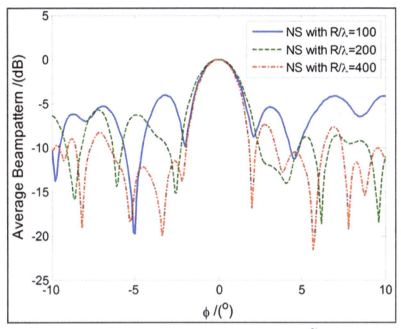

**Figure 16** Average beampattern of the selected CR nodes with $\tilde{R} = 100, 200, 400$

## 5 Conclusions

We have introduced the DB technique to the CR network, which is constituted of distributed CR nodes. The goal of the DB method is to forward the CR signal to the DCR user, while causing no harmful interferences to coexisting PUs by limiting its transmission power towards directions of PUs.

It is unavoidable that introducing the DB method to CR networks will lead to $\tilde{R}$ an extremely narrow main beam of the beampattern. To find a new network structure which has less impact of the working frequency, we have proposed a NS method. The NS method chooses those CR nodes, which are closely located to an end-fire array considering the direction of the DCR user. Our NS method can also be applied to cases with more than one DR users. The average beampattern of the proposed NS method show that the main beams are successfully directed towards the DCR users and are sufficiently enlarged for practical applications in CR networks. What is more, for a CR network with a large physical size, our NS method can widen the main beam while maintaining adequate low sidelobe levels for CR transmission.

## References

[1] J. Mitola, III, "Cognitive radio: an integrated agent architecture fro software defined radio," Ph.D., Royal institute of technology Stockholm, Sweden.

[2] S. Srinivasa and S. A. Jafar, "The throughput potential of cognitive radio: a theoretical perspective," in Signals, Systems and Computers, 2006. ACSSC '06. Fortieth Asilomar Conference on, 2006, pp. 221–225.

[3] O. Akan, O. Karli, and O. Ergul, "Cognitive radio sensor networks," Network, IEEE, vol. 23, pp. 34–40, 2009.

[4] H. Ochiai, P. Mitran, H. V. Poor, and V. Tarokh, "Collaborative beamforming for distributed wireless ad hoc sensor networks," Signal Processing, IEEE Transactions on, vol. 53, pp. 4110–4124, 2005.

[5] D. Lun, A. P. Petropulu, and H. V. Poor, "A cross-layer approach to collaborative beamforming for wireless Ad Hoc networks," Signal Processing, IEEE Transactions on, vol. 56, pp. 2981–2993, 2008.

[6] K. Zarifi, S. Affes, and A. Ghrayeb, "Distributed beamforming for wireless sensor networks with random node location," in Acoustics, Speech and Signal Processing, 2009. ICASSP 2009. IEEE International Conference on, 2009, pp. 2261–2264.

[7] R. Mudumbai, G. Barriac, and U. Madhow, "On the feasibility of distributed beamforming in wireless networks," Wireless Communications, IEEE Transactions on, vol. 6, pp. 1754–1763, 2007.

[8] J. Yindi and H. Jafarkhani, "Network beamforming using relays with perfect channel information," Information Theory, IEEE Transactions on, vol. 55, pp. 2499–2517, 2009.

[9] Z. Gan, W. Kai-Kit, A. Paulraj, and B. Ottersten, "Collaborative-relay beamforming with perfect CSI: optimum and distributed implementation," Signal Processing Letters, IEEE, vol. 16, pp. 257–260, 2009.

[10] V. Havary-Nassab, S. Shahbazpanahi, A. Grami, and L. Zhi-Quan, "Distributed beamforming for relay networks based on second-order statistics of the channel state information," Signal Processing, IEEE Transactions on, vol. 56, pp. 4306–4316, 2008.

[11] S. Fazeli-Dehkordy, S. Shahbazpanahi, and S. Gazor, "Multiple peer-to-peer communications using a network of relays," Signal Processing, IEEE Transactions on, vol. 57, pp. 3053–3062, 2009.

[12] C. Haihua, A. B. Gershman, and S. Shahbazpanahi, "Filter-and-Forward distributed beamforming in relay networks with frequency

selective fading," Signal Processing, IEEE Transactions on, vol. 58, pp. 1251–1262, 2010.

[13] D. R. Brown and H. V. Poor, "Time-slotted round-trip carrier synchronization for distributed beamforming," Signal Processing, IEEE Transactions on, vol. 56, pp. 5630–5643, 2008.

[14] R. Mudumbai, D. R. Brown, U. Madhow, and H. V. Poor, "Distributed transmit beamforming: challenges and recent progress," Communications Magazine, IEEE, vol. 47, pp. 102–110, 2009.

[15] X. Lian, H. Nikookar, and L. P. Ligthart, "Efficient radio transmission with adaptive and distributed beamforming for Intelligent WiMAX," Wirel. Pers. Commun., vol. 59, pp. 405–431, 2011.

[16] M. Jun, G. Y. Li, and J. Biing Hwang, "Signal processing in cognitive radio," Proceedings of the IEEE, vol. 97, pp. 805–823, 2009.

[17] K. Zarifi, S. Affes, and A. Ghrayeb, "Collaborative null-steering beamforming for uniformly distributed wireless sensor networks," Signal Processing, IEEE Transactions on, vol. 58, pp. 1889–1903, 2010.

[18] C. A. Balanis, Antenna Theory: Analysis and Design, 3rd Edition: Wiley-Interscience, 2005.

## Appendix A

*Proofs of Equations (16)–(21)*

We assume the Cartesian Coordinates of the $k$th CR node to be $(r_{k,x}, r_{k,y})$, where

$$r_{k,x} = r_k \cos \Psi_k \qquad (A - 1)$$

$$r_{k,y} = r_k \sin \Psi_k \qquad (A - 2)$$

Using equation (6), we can further separate the initial phase of the $k$th CR node into two parts, which are dedicated to the broadside array $\varphi_{k,b}$ and to the end-fire array $\varphi_{k,e}$. It means

$$\varphi_k = \varphi_{k,b} + \varphi_{k,e} \qquad (A - 3)$$

where

$$\varphi_{k,b} = -\frac{2\pi}{\lambda} r_k \sin \phi_0 \sin \Psi_k \qquad (A - 4)$$

$$\varphi_{k,e} = -\frac{2\pi}{\lambda} r_k \cos \phi_0 \cos \Psi_k \qquad (A - 5)$$

The array factor for the broadside array can be written as

$$
\begin{aligned}
F_{broadside}\left(\phi\right) &= \tfrac{1}{K} \sum_{k=1}^{K} \exp\left(j\tfrac{2\pi}{\lambda} r_k \sin \Psi_k \sin \phi + \varphi_{k,b}\right) \\
&= \tfrac{1}{K} \sum_{k=1}^{K} \exp\left[j\tfrac{2\pi}{\lambda} r_k \left(\sin \Psi_k \sin \phi - \sin \phi_0 \sin \Psi_k\right)\right] \qquad \text{(A - 6)} \\
&= \tfrac{1}{K} \sum_{k=1}^{K} \exp\left[j 2\pi \widetilde{R}(\sin \phi - \sin \phi_0)\tfrac{r_k}{R} \sin \Psi_k\right]
\end{aligned}
$$

Using the pdf of $z_k$ in equation (10), and the results in equations (10)–(13), we can derive in a similar way that

$$
\overline{P}_{broadside}(\phi) = E\left[|F\left(\phi\right)|^2\right] = \frac{1}{K} + \left(1 - \frac{1}{K}\right)\mu_b(\phi) \qquad \text{(A - 7)}
$$

where

$$
\mu_b(\phi) = \left|\frac{2J_1\left(\alpha_b\left(\phi\right)\right)}{\alpha_b\left(\phi\right)}\right| \qquad \text{(A - 8)}
$$

$$
\alpha_b\left(\phi\right) = 2\pi \widetilde{R}\left(\sin \phi - \sin \phi_0\right) \qquad \text{(A - 9)}
$$

Similarly we can also obtain the average beampattern for the end-fire array, which can be written as

$$
\overline{P}_{end-fire}(\phi) = \frac{1}{K} + \left(1 - \frac{1}{K}\right)\mu_e^2\left(\phi\right) \qquad \text{(A - 10)}
$$

where

$$
\mu_e(\phi) = \left|\frac{2J_1\left(\alpha_e\left(\phi\right)\right)}{\alpha_e\left(\phi\right)}\right| \qquad \text{(A - 11)}
$$

$$
\alpha_e\left(\phi\right) = 2\pi \widetilde{R}\left(\cos \phi - \cos \phi_0\right) \qquad \text{(A - 12)}
$$

The above equations (A-7)–(A-12) show the result of equations (16)–(21).

## Biographies

**Xiaohua Lian** was born on May 10 1980 in Urmqi, P.R. China. She received her Bachelor and Master of Engineering respectively in June 2002 and April 2005 from Nanjing University of Aeronautics and Astronautics, China. In Oct. 2013 she received her PhD degree from Delft University of Technology, The Netherlands. Her areas of interest include Smart antennas, Beamforming and cognitive radio.

**Homayoun Nikookar** is an Associate Professor in the Microwave Sensing Systems and Signals Group of Faculty of Electrical Engineering, Mathematics and Computer Science at Delft University of Technology. He has received several paper awards at international conferences and symposiums and the 'Supervisor of the Year Award' at Delft University of Technology in 2010. He is the Secretary of the scientific society on Communication, Navigation, Sensing and Services (CONASENSE). He has published more than 130 refereed journal and conference papers, coauthored a textbook on 'Introduction to Ultra

Wideband for Wireless Communications', Springer 2009, and has authored the book 'Wavelet Radio', Cambridge University Press, 2013.

**Leo P. Ligthart** was born in Rotterdam, the Netherlands, on September 15, 1946. He received an Engineer's degree (cum laude) and a Doctor of Technology degree from Delft University of Technology in 1969 and 1985, respectively. He received honorary Doctorates from Moscow State Technical University of Civil Aviation in 1999, Tomsk State University of Control Systems and Radioelectronics in 2001 and Military Technical Academy, Romania in 2010. He is an academician of the Russian Academy of Transport. From 1992 to 2010 he held the chair of Microwave Transmission, Radar and Remote Sensing in the Faculty of Electrical Engineering, Mathematics and Computer Science, Delft University of Technology. In 1994, he founded the International Research Center for Telecommunications and Radar (IRCTR) and was the director of IRCTR for more than 16 years. His principal areas of specialization include antennas and propagation, radar and remote sensing, but he has also been active in satellite, mobile and radio communications. He has published over 600 papers and book chapters, and 2 books. Prof. Ligthart is a Fellow of the IET, IEEE and is the Chairman of the CONASENSE Foundation.

# Wireless Sensor and Communication Nodes with Energy Harvesting

Mehmet Şafak

*Hacettepe University, Ankara, Turkey, E-mail: msafak@ee.hacettepe.edu.tr*

Received September 2013; Accepted November 2013
Publication January 2014

## Abstract

Energy is a valuable resource in wireless communication, navigation and sensor nodes. Maintenance and replacement of batteries in battery-driven nodes may not be possible, cost-effective or suitable for many applications. On the other hand, energy harvesting provides sustainable and independent operation with very long life-time but usually with unregulated power flow. The design approaches in present systems based on a regulated flow of power from an energy storage device are not optimal for energy harvesting nodes. This paper briefly reviews the capabilities of the present energy harvesting technologies and some intelligent design approaches based on unregulated supply power to energy harvesting nodes. The paper also presents a brief review of the alternative approaches for in-body communications with energy harvesting. The use of THz band does not seem to be feasible because of the excessive signal attenuation in the in-body channel. However, radio-frequency identification (RFID) with inductive magnetic coupling looks appropriate due to the fact that the human body behaves like free space to the magnetic field but strongly attenuates the electric field.

**Keywords:** Energy harvesting, wireless sensor networks, wireless communications, green communications, near-field communications, RFID, nanogenerators, THz communications, biological communication channels.

*Journal of Communication, Navigation, Sensing and Services, Vol. 1,* 47–66.
doi: 10.13052/jconasense2246-2120.113

# 1 Introduction

Energy is a valuable resource in communications, navigation, sensing and services (CONASENSE)-related applications. Operation of, for example, wireless communications and sensor networks (WSN) requires a regulated flow of power from the energy/power source to the electronic equipment. If electricity is not available on-site, the systems may be operated by battery or harvesting energy from their close vicinity. However, irrespective of whether the required electric energy is provided by the mains, the battery or energy harvesting, minimization of the consumed energy is strongly desired because of reasons such as cost, equipment life-time and electromagnetic compatibility. In WSN, sensors are usually battery-operated because of the difficulty and/or inconveniance of reaching sensor nodes in remote locations, high cost of maintainence and replacement. Hence, the energy efficiency of wireless sensor nodes determines the life-time of battery-operated sensor nodes, which are required to provide independent, sustainable and continuous operation. The design of new generation systems should take into account of the limitations of energy harvesting, e.g., scarcity, unregulated flow and non-availability of power in some time intervals which can not be predicted beforehand.

As shown in Figure 1, a communication or a sensor node may be considered to be composed of supply and demand sides. The *demand side* consists of energy consuming units such as a sensor, a signal processing unit, a wireless transceiver and a buffer, either to store the sensed data and/or for the data to be transmitted/received [1]. A sensor node differs from a communication node by the presence of a sensor. Transceivers typically use Bluetooth or Zigbee protocols to communicate within a range of maximum 30 m and require output power levels in the order of 2–100 mW (see Table 1) [2]. Hence, power levels needed by a sensor node may be in the order of a few 100 mWs. However, the energy consumption in transceivers may be decreased by reducing the data to be transmitted/received, chosing adaptive coding and modulation strategies, using energy efficient transmission scheduling, routing and medium access control as well as exploiting power saving modes (sleep/listen) [3], [4].

The *supply side* of an energy harvesting node consists of energy storage and energy harvesting systems. Energy harvesting implies the collection of energy from ambient sources and converting it into electrical energy. On the other hand, the lifetime of energy harvesting systems is theoretically infinite. However, battery-operated nodes do not have an energy harvester and their life-time is mainly limited by the battery capacity.

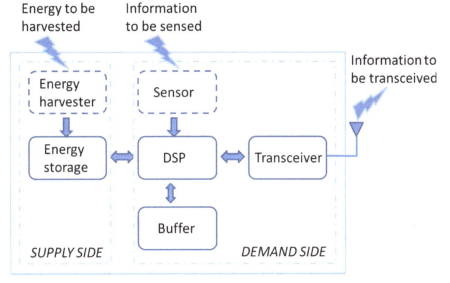

**Figure 1**   Block diagram of a sensor/communication node from energy perspective [3].

**Table 1**   Required output power and range of Bluetooth and Zigbee [2].

| Transceiver | Frequency | Bit Rate | Output Power | Range |
|---|---|---|---|---|
| IEEE 802.15.1 (Bluetooth) | ISM band(2.4 GHz) | 1 Mbps | 100 mW | <30m |
| IEEE 802.15.4 (Zigbee) | ISM band(2.4 GHz) | 250 kbps | 2 mW | <30m |

Recently remarkable improvements are observed in power density (W/kg), efficiency, amount of supplied power and the capacity (Amp.-hour) in the area of energy storage [5]. Nevertheless, operation by batteries still has its limitations and may not be suitable for certain applications. Hence, there is a strong demand for energy harvesting systems which can generate their own energy from their environment. Since energy to be harvested may not always be available and predictable, energy harvesting systems usually employ batteries for storing the harvested energy for present/future use. The harvesting efficiency and the availability of energy source are the fundamental issues to be considered. Since existing node designs are mostly based on the continuous flow and availability of the electric energy, these nodes-may not operate optimally with energy harvesting and novel approaches are required.

The next section of this paper will briefly review the capabilities offered by battery operation and energy harvesting. Section 3 will provide a brief discussion of intelligent node designs based on energy harvesting. Electromagnetic wave propagation in biological tissues in various frequency bands will be reviewed in Section 4 in order to discover the potential and the limitations of in-body communications. Section 5 will present near-field communications, which is based on wave-propagation in the near-field (Fresnel) region, and its use especially for in-body communications. Section 6 will provide a discussion of the potential, limitations of nanogenerators and their use in nano-sized sensor nodes operating in the THz band. Finally, the conclusions will be presented in Section 7.

## 2 Energy Supply

*Battery-driven systems* use stored chemical energy to supply the electrical energy needed by a node. Hence the batteries and the battery-driven systems have a finite lifetime. It is noted that regular maintenance and replacement of batteries may be difficult and costly when nodes are remotely located and/or densely populated. To increase the life time of battery-driven nodes, one may use either higher battery capacity, which implies increased cost, or low-duty cycle operation with lower sensing reliability. Increased transmission range requires higher transmit power levels, while lower transmission range (lowering transmit power) implies more hops and usually higher energy usage at multiple nodes. If both demand and supply sides of a node as shown in Figure 1 are optimized, the energy need may be decreased by an order of magnitude compared to present designs; then, the energy need of the demand side may be reduced to the order of several 10 mW [6].

Table 2 presents the results of a recent survey of the energy levels that can be supplied by various battery types. One may easily observe from Table 2 that even if the energy demand of a node is decreased by an order of magnitude with intelligent designs, battery-operated systems sustain power levels in the order of several 10 mW only for limited time durations, hence limited life-time.

*Energy harvesting* requires both an energy source to be harvested and an energy harvesting sensor, which should be matched to the energy source for optimizing the harvesting process. The energy sources to be harvested may be thermal (including solar energy and human body), mechanical (translational/vibrational), chemical, electromagnetic, wind etc. One may use piezoelectric sensors (vibrational energy), thermocouples (thermal energy), photovoltaic cells (solar energy), wind turbines (wind energy) and

**Table 2** Present technology for battery power [5].

| Battery Type | Nominal Voltage (V) | Capacity Capacity | Weight Energy Density (Wh/kg) | Power Density (W/kg) | Efficiency (%) | Self Discharge (%month) |
|---|---|---|---|---|---|---|
| SLA | 6 | 1300 | 26 | 180 | 70–92 | 20 |
| NiCd | 1.2 | 1100 | 42 | 150 | 70–90 | 10 |
| NiMH | 1.2 | 2500 | 100 | 250–1000 | 66 | 20 |
| Li-ion | 3.7 | 740 | 165 | 1800 | 99.9 | <10 |
| Li-polymer | 3.7 | 930 | 156 | 3000 | 99.8 | <10 |

radio-frequency (RF) systems (electromagnetic energy) to harvest various energy types.

Limitations in energy harvesting usually come from uncontrollability and unpredictability of the energy source and the efficiency with which the harvested energy is converted into electrical energy. Consequently, the level of the harvested energy and its management (harvesting architecture) is critical in the system operation. In that context, storing or no-storing of the harvested energy and the optimal design of the demand side are serious issues to be considered. Energy usage should be carefully coordinated and optimized between supply and demand sides, including signal processing and communication architectures, sleep scheduling, energy-efficient communication protocols and adaptive coding and modulation [3].

*RF energy harvesting* is mainly based on magnetic inductive coupling due to the Faraday's law, which also provides the basic principles of electric motors and generators. A loop carrying a (sufficiently-fast) time-varying current, creates a magnetic field around itself. This induces an open-circuit voltage around the terminals of a receive loop, when it is brought to the vicinity of the transmit loop. The open-circuit voltage is proportional to the magnetic flux through the receive loop. In addition to the frequency and the amplitude of the transmit loop current, the dimensions, relative directions, the distance between and the number of turns of transmit and receive loops determine the flux and hence the open-circuit voltage. The two loops are required to be in the Fresnel region of each other, where radiative near-field communications takes place. The voltage induced at the terminals of the receive loop may be used either for responding the request of the transmitting loop as in RFID systems, or for charging a battery connected to its terminals. Induced voltage levels in these systems are reported to be approximately 0.5 V [5].

*Vibrational energy harvesters,* made of piezoelectric material, convert vibrational (mechanical) energy into electrical energy. An electric potential is induced at the terminals of a piezoelectric material if it undergoes a strain (e. g., due to body motion, muscle stretching, breathing, sonic waves) due to the polarization of ions in the crystal. This electric potential may drive a transient flow of electrons in the external circuit, hence it generates electric energy; this is the fundamental principle of the nanogenerator. It can as well cause the flow of charge carriers through a semiconductor material, resulting in piezopotential-gated field-effect-transistors, diodes and sensors; this is the so-called principle of piezotronics [7] [8]. Piezoelectric energy harvesters directly convert a mechanical vibration into a relatively high voltage (~1–20 V) and output current (~1–100 $\mu$A) with a duty cycle less than 3 % in RF transmissions. Piezoelectric energy harvesters are used in railway and road tunnels to produce electricity by using the vibrations induced by trains and cars. The applications also include shoe-powered RF tag system, which converts the vibrational energy created by walking subjects, and self-powered door bells, which produce electrical energy using the vibrations when a door bell is pressed [9]. Electrical power may also be generated for applications in bio-microelectromechanical systems (MEMS) and microelectronic systems by inserting piezoelectric generators between the coping of a dental crown and the metal abutment [10].

*Thermal energy* (TE) *generators* convert temperature differences into electrical energy based on the so-called Seebeck effect which was first discovered in 1821 by T. J. Seebeck. A simple thermal energy generator may be made by heating one face of a TE module and cooling the other face, causing an electric current flowing through a load connected to its terminals. A TE generator has usually a long life cycle, no moving parts, simple and high reliability but low-efficiency (5–10%). Among the thermal energy sources, one may list the waste heat in industrial applications and solar thermal energy. Thermal energy can also be harvested from a human body exploiting small temperature gradients. Seiko thermic watch harvests 22 $\mu$W from a human body and uses this power to drive the wrist watch and charges a 4.5 mAh lithium-ion battery. When it is worn on the wrist, the watch uses the body heat absorbed from its back case to generate power with its thermal converter. The power generating capability is a function of the difference between the air temperature and the body surface temperature. If the air temperature is equal to, or greater than, the surface temperature of the body, the watch is unable to generate power [11]. Human body has potential in harvesting thermal and kinetic energies. The harvested power depends on whether the subject is walking or running. For example, it

**Table 3** Energy sources and their energy harvesting potential.

| Energy Source | Available Power | Conversion Efficiency | Harvested Power |
|---|---|---|---|
| Solar [4] | 100 mW/cm$^2$ | 15% | 15mW/cm$^2$ |
| Wind [4] | – | – | 120mWh/day |
| Finger motion [4] | 19mW | 11% | 2.1mW |
| Footfalls [4] | 67W | 7.5% | 5W |
| Exhalation [4] | 1W | 40% | 0.4W |
| Breathing [4] | 0.83W | 50% | 0.42W |
| Blood pressure [4] | 0.93W | 40% | 0.37W |
| Thermoelectric generator [14] | – | – | 3.5mW/cm$^2$ @30$^0$C gradient |
| Human body (walking subject) [10] | – | – | 6$\mu$W/cm$^3$(thermal) 1.3$\mu$W/cm$^3$(kinetic) |
| Inductive magnetic coupling [15] | – | 73% @5cm 24%@15cm | – |

is reported in [12] that one can harvest approximately 6 $\mu$W thermal energy and a kinetic energy of 1.3 $\mu$W/cm$^3$ from the body of a walking subject. The corresponding numbers are 10 $\mu$W and ∼35$\mu$W for a running subject.

The human body may be used in various other ways to harvest energy, e.g., via blood pressure and breath, which are uncontrollable by the user, and via finger motion, paddling (bycle dynamo) and walking (shoes), which are user controllable. Wearable bio-sensors such as gloves, wrist-watches, rings, patches, earlobes, intelligent clothes, eye-glasses, accelerometers, glucose monitor, electrocardiograph, pulse sensor, oxygen-level monitor, temperature sensor, respiratory meter can potentially be used for harvesting energy from human body [13].

Table 3 shows a summary of some energy sources, their characteristics, required energy harvesting technology and the amount of energy that can potentially be harvested.

## 3 Intelligent Designs for Energy Harvesting Nodes

Classical design of sensor/communication nodes is based on the availability of a continuous flow of a constant power level (infinite energy) to the demand side. On the other hand, for the self-powered nodes to be practical, dramatic reductions are strongly desired in the dissipated power levels since energy harvesting technology is presently far from satisfying present needs. Energy sources may be (un)controllable and/or (un)predictable for energy harvesting; solar energy is predictable but uncontrollable, while RF energy harvesting in an RFID system may be controllable and predictable at the same time. Therefore, limited power that can be harvested sets a constraint on the average power consumed by the demand side for self-powered operation. This implies that energy harvesting, storing and processing technologies should be improved so as to help sustainable and continuous operation.

Even if infinite energy becomes available to the supply side, energy generation may not be continuous and/or rate of generation may be limited. Storing the harvested energy may partially alleviate this problem since it may regulate the power flow. Nevertheless, electronic devices with classical design can not reliably operate under these conditions. Therefore, energy generation profile of the supply side should be matched to the energy consumption profile of the demand side. This requires a system-level approach involving variation-tolerant architectures, ultra-low voltage levels and highly digital RF circuits. In addition, one needs DSP architecture and circuits which are energy-efficient, energy-scalable, and robust to variations in the output voltage/power levels of the supply side. Energy-scalable hardware may call for techniques for approximate processing, which implies a trade-off between power and arithmetic precision [16]. In wireless sensor networks, the demand side may be designed with sleep/awake periods in synchronism with energy harvesting by the supply side. Energy consumption policy may be optimized in seeking a tradeoff between the throughput and the life-time of the sensor node [17]. Such approaches are believed to result in more than an order of magnitude energy reduction compared to present systems [6].

In some projects like PicoRadio (Berkeley), $\mu$AMPS (MIT), WSSN (ICT Vienna) and GAP4S (UT Dallas), densely populated low-cost sensor nodes are foreseen to operate with power levels of approximately 100 $\mu$W; such power level is believed to be within the capabilities of energy harvesting [18]. Even though dramatic improvements are still needed in the energy harvesting technologies for self-operated nodes, rapidly-evolving energy harvesting technologies are believed to be promising.

# 4  Electromagnetic Wave Propagation in Biological Tissues

An electric field propagating in a lossy dielectric medium may be written as

$$E = E_0 \, e^{-\gamma r} = E_0 \, e^{-\alpha r} \, e^{-j\beta r} \tag{1}$$

where $\gamma = \alpha + j\beta$ denotes the complex propagation constant, $r$ is the distance and $E_0$ is the value of the electric field at $r = 0$. In a lossy dielectric medium, the complex propagation constant may be written as [19]

$$\gamma = \alpha + j\beta = k_0\sqrt{\varepsilon_r - jp}$$

$$\alpha = k_0\sqrt{\frac{\sqrt{\varepsilon_r^2 + p^2} - \varepsilon_r}{2}}$$

$$\beta = k_0\sqrt{\frac{\sqrt{\varepsilon_r^2 + p^2} + \varepsilon_r}{2}} \tag{2}$$

$$p = \frac{\sigma}{w\varepsilon_0} = \frac{18\,\sigma}{f_{GHz}}$$

where $w$ and $k_0$ denote respectively the radial frequency and the free-space wave-number. The relative dielectric constant (permittivity) $\epsilon_r$ and the conductivity of the lossy dielectric medium $\sigma$ are both assumed to be independent of frequency. Note that, in a non-conducting medium ($\sigma=0$), the attenuation constant $\alpha$ vanishes and $\beta = k_0\sqrt{\varepsilon_r}$, as expected. In a lossy dielectric medium, $\sigma$ is related to the dispacement current. The signal attenuation in dB at a distance $r$ is given by

$$20 \log\left(e^{-\alpha r}\right) = 8.68\,\alpha\,r \quad (dB) \tag{3}$$

where $8.68\alpha$ (dB/m) denotes the attenuation coefficient.

In biological tissues, the dielectric constant is a frequency- and temperature-dependent complex quantity and provides a measure of the interaction of electromagnetic waves with tissue constituents at cellular and molecular levels [20], [21]. There is a lag (delay) between changes in polarization in the tissue and time-changes in the applied electric field due to the relaxation process. Consequently, the permittivity of a dielectric material becomes a complicated, complex-valued function of the frequency, which implies frequency-dependent and delayed response of live tissues to wave propagation. This dispersion effect is analogous to hysteresis in changing magnetic fields.

Frequency dependence of the relative dielectric constant may be modeled by the Cole-Cole equation:

$$\varepsilon_r = \varepsilon_\infty + \frac{\varepsilon_s - \varepsilon_\infty}{1 + (jw\tau)^{1-\alpha}} \tag{4}$$

which reduces to the so-called Debye equation for $\alpha=0$. Here, $\epsilon_s$ and $\epsilon_\infty$ denote respectively the values of the relative dielectric constant for $w\tau << 1$ (low-frequencies) and $w\tau >> 1$ (high frequencies). The static value $\epsilon_s$ of the relative dielectric constant is proportional to the water content of the tissue. The value of $\alpha$ is equal to zero for pure water and negligibly small for body fluids but greater than zero for most tissues. Here, $\tau$ denotes the mean relaxation time, which is longer than that for pure water.

An approximation to the complex dielectric constant given by (4) is given by [22]

$$\varepsilon_r \cong \varepsilon_\infty + \frac{\beta_0 + jw\beta_1}{\alpha_0 - w^2 + jw\alpha_1} \tag{5}$$

where $\alpha_0, \beta_0, \alpha_1, \beta_1$ are fitting parameters to the measured data. Inserting the value of the complex relative dielectric constant given by (4) or (5) into (2), one can easily find the corresponding values of the coefficients, $\alpha$ and $\beta$, of the complex propagation constant.

The relative magnetic permeability of a biological tissue in the RF band of interest is equal to unity ($\mu_r = 1$). Therefore, biological tissues may be characterized electrically by the relative dielectric constant and the conductivity only. A comprehensive experimental study [20] provides in-vivo measurement values of these two parameters of some animal tissues at frequencies varying between 50 MHz and 20 GHz. Some results for human tissues are also given. The provided data may be used for modeling the in-body communication channel, e.g., for signal transmission between implant-devices and devices on/outside the body.

Table 4 shows the measurement results for relative dielectric constant and conductivity of some in-vivo body tissues of $\sim$50 kg pigs at 50 MHz, 1 GHz and 20 GHz. The corresponding attenuation coefficients in dB/cm are also provided. Table 4 clearly shows that, for electromagnetic wave propagation in biological tissues, the attenuation coefficient expressed in dB/cm is acceptably low at 50 MHz but increases to relatively high values at 1 GHz and becomes excessively high at 20 GHz. On the other hand, in view of the typical transmission ranges in human body, far-field communications is not possible at 50 MHz (the wavelenth $\lambda = 6$m) since transmit and receive antennas may not be in the far-fields of each other. At 1 GHz ($\lambda = 0.3$m), although the transmitter and receiver may be located in far-field regions of each other, very high transmit powers are required for compensating excessively high channel attenuations. Such high transmit powers are also unacceptable for health reasons.

In addition to the very high attenuation coefficients for in-body communications, as given by Table 4, one should also consider the free-space path loss, $(4\pi r/\lambda)^2$, which is proportional to $r^2$ and $f^2$. Also note that

**Table 4** Relative permittivity, conductivity and attenuation coefficient of biological tissues at 50 MHz, 1 GHz and 100 GHz (based on [20]*).

| Tissue | Parameter | 50 MHz | 1 GHz | 20 GHz |
|---|---|---|---|---|
| Liver | $\epsilon_r$ | 96 | 50 | 30 |
| | $\sigma$(S/m) | 0.7 | 1 | 20 |
| | 8.68$\alpha$ (dB/cm) | 0.85 | 2.28 | 57.41 |
| Lung (deflated) | $\epsilon_r$ | 87 | 40 | 26 |
| | $\sigma$(S/m) | 0.8 | 1 | 17 |
| | 8.68$\alpha$ (dB/cm) | 0.94 | 2.53 | 52.49 |
| Long bone | $\epsilon_r$ | 43 | 25 | 15 |
| | $\sigma$(S/m) | 0.2 | 0.4 | 8 |
| | 8.68$\alpha$ (dB/cm) | 0.41 | 1.3 | 32.9 |
| Bone marrow (50% in-vivo) | $\epsilon_r$ | 67 | 35 | 22 |
| | $\sigma$(S/m) | 0.6 | 0.8 | 14 |
| | 8.68$\alpha$ (dB/cm) | 0.81 | 2.17 | 47.1 |
| Fat (infiltrated) | $\epsilon_r$ | 16.5 | 15.5 | 9.0 |
| | $\sigma$(S/m) | 0.25 | 0.4 | 6.0 |
| | 8.68$\alpha$ (dB/cm) | 0.56 | 1.62 | 31.44 |
| Skin (in-vivo) | $\epsilon_r$ | 90 | 38 | 22 |
| | $\sigma$(S/m) | 0.2 | 0.6 | 13 |
| | 8.68$\alpha$ (dB/cm) | 0.32 | 1.58 | 43.9 |
| Vitreous humour | $\epsilon_r$ | 105 | 69 | 40 |
| | $\sigma$(S/m) | 1.8 | 1.9 | 60 |
| | 8.68$\alpha$ (dB/cm) | 1.5 | 3.64 | 134.1 |
| Cerebrospinal fluid (CSF) | $\epsilon_r$ | 105 | 70 | 44 |
| | $\sigma$(S/m) | 1.8 | 1.9 | 60 |
| | 8.68$\alpha$ (dB/cm) | 1.51 | 3.62 | 130.2 |

*Tabulated values are obtained by reading the curves provided in [20] and hence may contain some reading errors.

multi-path propagation, shadowing and scattering effects due to inhomogeneities in the electrical parameters of the tissues make the in-body communications even more challenging.

## 5 Near-Field Communications

Near-field communications refers to communications in the Fresnel region, defined by the range $0.62\sqrt{D^3/\lambda} < r < 2D^2/\lambda$, where $D$ denotes the largest antenna dimension. In this region, the relation between electric and magnetic fields radiated by an antenna is not as simple as in the far-field (Fraunhofer) region, which is defined by $r > 2D^2/\lambda$. In the Fresnel region, electric and magnetic fields do not have plane wavefronts and the ratio of electric to magnetic field intensities is not equal to the free-space intrinsic impedance, $120\pi$ Ohm, as in the far-field region. Nevertheless, the radiative field components still dominate the reactive field components and are used for near-field communications. Note that near- and far-field distances defined above are meaningful for antenna sizes exceeding the wavelength $(D>\lambda)$. For electrically small antennas with $D<\lambda/2$, the Fresnel region is usually assumed to be bounded by the range $\lambda/(2\pi) <r< \lambda$ [23].

The efficiency of near-field communications is critically dependent on coupling between transmit and receive antennas which are located in each other's Fresnel regions. Depending on the frequency of operation, the propagation medium and the types of antennas used, it might be more appropriate to use electric or magnetic coupling between antennas. The electrical characteristics (permittivity, magnetic permeability and conductivity) of the medium between transmit and receive antennas also plays a critical role in choosing the type of coupling. For example, a near-field communication system between, for example, implant devices inside a human body and an antenna over the skin should prefer magnetic coupling since the propagation medium (human body) is characterized by high conductivity and high relative dielectric constants (see Table 4). Since electric field suffers much higher attenuations in the in-body channel, magnetic coupling between transmit and receive antennas is more appropriate than electrical coupling. Since $\mu_r=1$ in the human body, magnetic fields do not suffer losses in addition to the free-space loss.

RFID systems constitute one of the most notable examples of inductive magnetic coupling which is commonly used in near-field communications [24]. The magnetic field radiated by time-varying currents in the RFID reader loop antenna induces a voltage around the terminals of the loop antenna of the RFID tag. The induced voltage is used for transmitting the information requested by the RFID reader. The efficiency of coupling between transmitter and receiver, hence the induced voltage, is determined by the fraction of the magnetic flux created by the transmit loop antenna coupled into the receive (RFID tag) loop antenna. Near-field coupling between antennas is highly dependent on the types, electrical sizes, radiation patterns, relative orientations and impedance matching of the antennas employed and the distance between them [25].

The same principle also applies for distant charging of a battery connected at the terminals of the receive loop antenna. For example, batteries of implant devices may be charged by inductive magnetic coupling. This also provides an example of RF energy harvesting, since received RF energy is converted into chemical energy in an in-body battery. However, it is reported that inductive charging efficiency is lower (hence slower charging) and resistive heating is higher [15]. RF wireless charging of multiple sensors is also considered for smart grid applications by using mobile chargers [26].

A recent study on wireless power transfer, based on strongly coupled magnetic resonance in the near-field, reports coupling coefficients as high as 0.7–0.9 between primary and secondary loops, as compared to coupling coefficients on the order of 0.1 for inductive magnetic coupling. The distance between the two resonators may be larger than the characteristic sizes of each resonator and, unlike for conventional inductive coupling, energy dissipations are reported to be small [15].

## 6 Nanogenerators and Communications in THz Band

Recently there has been considerable interest in nano-sized sensing and communicating devices that can detect and measure events at nanoscale where energy consumption is believed to be low. Energy harvesting at nanoscale is believed to provide independent, sustainable, maintenance-free, continuous operation. Nanogenerators are foreseen to be used for a variety of applications including intra-body drug delivery, health monitoring, medical imaging, environmental research (air pollution control), military applications (surveillance networks against nuclear, biological, and chemical attacks at nanoscale, and home security), and very high data rate communications [27].

For example, in contrast with today's cancer drugs which kill healthy cells as well as the cancerous ones, drugs may be delivered locally to diseased cells and tumors by miniature bio-MEMS systems that navigate the circulatory system of the human body. However, these machines, which are difficult to recharge within the body, may potentially harvest mechanical energy from the blood cells [10]. Today's technology may not be mature enough to build a motor and power source small enough to squeeze inside the capilleries that feed the tumor. However, magnetic drug carriers as small as 50 $\mu$m can be steered by external magnetic resonance imaging (MRI) machines [28]. In 2011 it is demonstrated that a gentle straining can yield 1–3 V with an instantaneous power of $\sim 2\mu$W from an integrated nanogenerator sheet of 1 cm$^2$ using a self-powered nanosensor. Potential applications for MEMS may require power levels in the range $\mu$W to mW. Future of nanotechnology research is likely to focus on technologies that provide higher harvested power levels and integration of nanosensors into nanosystems acting like living species with sensing, communicating, controlling, navigating and reacting [29].

Communication between sensor nodes is considered to be achievable in the terahertz (THz) band, 0.3–3 THz (1 mm $\geq \lambda \geq$ 0.1 mm). The radiation in the THz band is non-ionizing and covers part of the infrared (IR) band of the solar spectrum. THz band is currently used and/or considered for use in applications including radioastronomy, space-remote sensing, through-the-wall imaging, and medical surface imaging (for skin cancer).

As for communications in the THz band, the following issues need to be taken into account. Atmospheric absorption exceeds 100 dB/km and rain attenuation exceeds 3 dB/km and 10 dB/km for rain rates of 5 mm/hr and 25 mm/hr, respectively, at 1 THz. On the other hand, free-space propagation loss is equal to 152.4 dB at a distance of 1 km at 1 THz. This limits the applicability of communications in the THz band only to short ranges. Since far-field distance is on the order of a fraction of a millimeter, near-field communications is not possible in the THz band. It is noted that, in view of Table 4, for communications between nano-sensors inside a human body and a receiver on-body, the communications budget is not better-off because of the excessively large attenuation. For example, electrical parameters of a deflated lung at 20 GHz are measured to be $\epsilon_r = 26$, $\sigma = 17$ S/m (see Table 4); the corresponding attenuation constant is 52.49 dB/cm. Assuming that these values of the electrical parameters at 20 GHz are also valid at 100 GHz, which looks to be a reasonable assumption [19], the attenuation coefficient is found using (3) to be 8.68 $\alpha = 54.45$ dB/cm at 100 GHz. Hence, the total attenuation at a typical distance of 15 cm becomes 828.9 dB including the free-space propagation loss of 55.96 dB. In contrast with this, the attenuation through 15 cm thickness of biological tissue is measured to be 47 dB and 39 dB at 13.5 MHz and 35 MHz, respectively, in an inductive coupling near-field communication system [30]. Multipath propagation, shadowing and scattering due to inhomogeneities in the human body may render in-body communication seven more difficult. Transmit power levels required to compensate for such large losses are unlikely to be obtained by energy harvesting. On the other hand, the use of electrically large antennas will lead to problems due to pointing errors.

In summary, due to the unavailability of THz sources and of compact, solid-state, room-temperature transceivers, we do not foresee THz commmunications as a good option in near-to mid-terms [31]. However, nanogenerators and nanopiezotronics, which refers to the technology based on coupling piezoelectric and electronics properties, are listed among the top 10 emerging technologies for the future [32]-[34].

## 7 Conclusions

With ever-increasing needs for mobility and higher data rates, energy becomes a limiting factor for the performance, the sustainability and the life-time of wireless communication, navigation and sensor nodes. On the other hand, maintenance and replacement of batteries in battery-driven nodes may not be possible, cost-effective or appropriate in many applications. Energy harvesting may provide sustainable and independent operation with very long life-times and is well-suited for some applications such as sensing. However, this technology, presently in its infancy, can not yet meet the requirements for regulated power/energy levels. Present design approaches for nodes in CONASENSE applications, which are mostly based on the supply of regulated power levels from energy storing devices, are therefore not optimal for energy harvesting purposes. In view of limited and noncontinuous availability of the energy to be harvested at varying generation rates, the design of the demand (energy consuming) side of a node should be matched to the supply side, which consists of energy harvesting and storing units. Efficient node designs should therefore be based on low-power consumption and energy-scalability. For in-body communications with energy harvesting, near-field communications with inductive magnetic coupling seem to be appropriate because the human body behaves like free space to the magnetic field but strongly attenuates the electric field. Micro- and nano-systems may be promising for energy harvesting applications in mid- to far-terms. However, in-body communications in the THz band using micro- and nano-systems seems to be unlikely, mainly because of the excessive signal attenuation in the in-body channel.

## References

[1] Niyato, D., E. Hossain, M. M. Rashid and V. K. Bhargava, "Wireless sensor networks with energy harvesting technologies: a game-theoretic approach to optimal energy management," *IEEE Wireless Communications*, August 2007, pp. 90–96.
[2] Nakajima, N., Short-range wireless network and wearable bio-sensors for healthcare applications, 2nd Int. Symposium on Applied Sciences in Biomedical and Communication Technologies (ISABEL 2009), 2009, pp.1–6
[3] V. Sharma, U. Mukherji, V. Joseph and S. Gupta, "Optimal energy management policies for energy harvesting sensor nodes," *IEEE Trans. Wireless Communications*, vol. 9, no.4, pp.1326–1336, April 2010.

[4] Joseph, V., V. Sharma, and U. Mukherji, "Optimal sleep-wake policies for an energy harvesting sensor node," *IEEE ICC*, 2009.

[5] Sudevalayam, S., and P. Kulkarni, "Energy Harvesting Sensor Nodes: Surveys and Implications," *IEEE Communications Surveys & Tutorials*, vol.13, no.3, 3rd Quarter 2011.

[6] Chandrakasan, A. P., D. C. Daly, J. Kwong and Y. K. Ramadass, "Next-generation micro-power systems," *IEEE Symposium on VLSI circuits, Digest of technical papers*, 2008, pp.2–5.

[7] Wang, Z. L., "Top emerging technologies for self-powered nanosystems: nanogenerators and nanopiezotronics," *3rd Int. Nanoelectronics Conf. (INEC)*, pp.63–64, 2010.

[8] Chao, P. C.-P., "Energy harvesting electronics for vibratory devices in self-powered sensors, *IEEE Sensors Journal*, vol. 11, no.12, pp.3106–3121, December 2011.

[9] Kroener, M., "Energy harvesting technologies: energy sources, generators and management for wireless autonomous applications," *9th Int. Multi-Conf. Systems, Signals and Devices (SSD)*, 2012, pp.1–4.

[10] Mhetre, M. R., N. S. Nagdeo, and H. K. Abhyankar, "Micro energy harvesting for biomedical applications: a review," *3rd Int. Conf. Electronics Computer Technology (ICECT)*, 2011, vol.3, pp.1–5.

[11] Lu, X., and S.-H. Yang, "Thermal energy harvesting for WSNs,"*IEEE Int. Conf. Systems, Man & Cybernatics (SMC)*, 2010, pp.3045–3052.

[12] Mitcheson, P. D.,"Energy harvesting for human wearable and implantable bio-sensors," *Annual Int. Conf. IEEE Eng. Medicine and Biology Society (EMBS)*, 2010, pp.3432–3436.

[13] Teng, X. F., Y. T. Zhang, C. C. Y. Poon, and P. Bonato, "Wearable medical systems for p-health, " *IEEE Reviews in Biomedical Engineering*, vol.1, 2008, pp. 62–74.

[14] Wan, Z. G., Y. K. Tan, and C. Yuen, "Review on energy harvesting and energy management for sustainable wireless sensor networks," *IEEE 13$^{th}$ Int. Conf. on Communication Technologies (ICCT)*, pp. 362–367, 2011.

[15] R. Amirtharajah, J. Collier, J. Siebert, B. Zhou, and A. Chandrakasan, "DSPs for energy harvesting sensors, Applications and Architectures," *IEEE Pervasive Computing*, July–Sept. 2005, pp. 72–79.

[16] Rajesh, R., V. Sharma and P. Viswanath, "Capacity of fading Gaussian channel with an energy harvesting sensor node," *IEEE Globecom Conf. 2011.*

[17] Tacca, M., P. Monti and A. Fumagalli, "Cooperative and reliable ARQ protocols for energy harvesting wireless sensor nodes," *IEEE Trans. Wireless Communications*, vol.6, no.7, pp. 2519–2529, July 2007.

[18] Ames, L. A., J. T. de Bettencourt, J. W. Frazier and A. S. Orange, "Radio Communications via Rock Strata," *IEEE Trans. Communication Systems*, 1963, pp.159–169.

[19] Peyman, A., S. Holden and C. Gabriel, "Dielectric properties of tissues at microwave frequencies," Report, Mobile Telecommunications and Health Research Programme, December 2009.http://www.mthr.org.uk /research_projects/documents/Rum3FinalReport.pdf.

[20] Gabriel, C., S. Gabriel, and E. Corthout, "The dielectric properties of biological tissues: I. Literature survey," Phys. Med. Biol., vol.41, 1996, pp.2231–2249.

[21] Khaleghi, A., I. Balasingham, and R. Chavez-Santiago, "Computational study of the ultra-wideband wave propagation into the human chest," *IET Microwaves, Antennas and Propagation*, vol. 5, no. 5, pp.559–567, 2011.

[22] Balanis, C. A., *Antenna Theory: Analysis and Design (3rd ed.)*, NJ: J. Wiley, 2005.

[23] Ortiz, S., Jr., "Is near-field communication close to success?," *Computer*, March 2006, pp.18–20

[24] Chen, Y. S., S.-Y. Chen and H.-J. Li, "Analysis of antenna coupling in near-field communication systems," *IEEE Trans. Antennas Propagation*, vol.58, no.10, pp.3327–3335, October 2010.

[25] Ho, S. L., J. Wang, W. N. Fu, and M. Sun, "A comparative study between novel witricity and traditional inductive magnetic coupling in wireless charging," *IEEE Trans. Magnetics*, vol.47, no.5, pp.1522–1525, May 2011.

[26] M. E. Kantarci, and H. T. Mouftah, "SuReSense: sustainable wireless rechargeable sensor networks for the smart grid," *IEEE Wireless Communications*, pp. 30–36, June 2012.

[27] Jornet, J. M., and I. F. Akyildiz, "Joint energy harvesting and communication analysis for perpetual wireless nanosensor networks in the Terahertz band," *IEEE Trans. Nanotechnology*, vol.11, no.3, pp.570–580, May 2012.

[28] Martel, S., "Journey to the center of a tumor," *IEEE Spectrum*, pp. 49–53, October 2012.

[29] Wang, Z. L.,"Nanogenerators for self-powering nanosystems and piezotronics for smart MEMS/NEMS," *IEEE 24th Int. Conf. MEMS*, pp. 115–120, 2011.

[30] T. Yamada, et al., "Battery-less wireless communication system through human body for in-vivo healthcare chip," *2004 Topical Meeting on Silicon Monolithic Integrated Circuits in RF Systems*, pp.322–325.

[31] Armstrong, C. M., "The truth about terahertz,"*IEEE Spectrum*, pp.28–33, Sept. 2012.

[32] Wang, Z. L., Top emerging technologies for self-powered nanosystems: nanogenerators and nanopiezotronics, *3rd Int. Nanoelectronics Conf. (INEC)*, 2010, pp.63–64.]

[33] New Scientists (Top 10 Future Technologies) http://www.newscientist.com/article/mg20126921.800-ten-scifi-devices-that-could-soon-be-in-your-hands.html?full=true] MIT Technology Review (Top 10 Emerging Technology in 2009), http://www.technologyreview.com/video/?vid=257

## Biography

**Mehmet Şafak** received the B.Sc. degree in Electrical Engineering from Middle East Technical University, Ankara, Turkey in 1970 and M.Sc. and Ph.D. degrees from Louvain University, Belgium in 1972 and 1975, respectively.

He joined the Department of Electrical and Electronics Engineering of Hacettepe University, Ankara, Turkey in 1975. He was a postdoctoral research fellow in Eindhoven University of Technology, The Netherlands during the academic year 1975–1976. From 1984 to 1992, he was with the Satellite Communications Division of NATO C3 Agency (formerly SHAPE Technical Centre), The Hague, The Netherlands, as a principal scientist. During this period, he was involved with various aspects of military SATCOM systems and represented NATO C3 Agency in various NATO committees and meetings.

In 1993, he joined the Department of Electrical and Electronics Engineering of Eastern Mediterranean University, North Cyprus, as a full professor and was the Chairman from October 1994 to March 1996. Since March 1996, he is with the Department of Electrical and Electronics Engineering of Hacettepe University, Ankara, Turkey, where he acted as the Department Chairman during 1998–2001. He is currently the Head of the Telecommunications Group.

He conducted and supervised projects, served as a consultant and organized courses for various companies and institutions on diverse civilian and military communication systems. He served as a member of the executive committee of TUBITAK (Turkish Scientific and Technical Research Council)'s group on electrical and electronics engineering and informatics. He acted as reviewer in various national and EU projects and for distinguished journals. He was involved in the technical programme committee of many national and international conferences. He served as the Chair of $19^{th}$ IEEE Conference on Signal Processing and Communications Applications (SIU 2011). He represented Turkey to COST Action 262 on Spread Spectrum Systems and Techniques in Wired and Wireless Communications. He acted as the chairman of the COST Action 289 Spectrum and Power Efficient Broadband Communications.

He was involved with high frequency asymptotic techniques, reflector antennas, wave propagation in disturbed SATCOM links, design and analysis of military SATCOM systems and spread spectrum communications. His recent research interests include multi-carrier communications, channel modelling, cooperative communications, cognitive radio and MIMO systems.

# ICT-based Remote Agro-Ecological Monitoring System – A Case Study in Taiwan

Cheng-Long Chuang[1,2] and Joe-Air Jiang[3,4,*]

[1]*Intel Labs, Intel Corporation,*
[2]*Intel-NTU Connected Context Computing Centre, National Taiwan University,*
*clchuang@ieee.org*
[3]*Department of Bio-Industrial Mechatronics Engineering, National Taiwan University,*
[4]*Education and Research Centre for Bio-Industrial Automation, National Taiwan University, jajiang@ntu.edu.tw*

Received September 2013; Accepted November 2013
Publication January 2014

## Abstract

In recent years, information and communication technologies have opened many opportunities to modern agriculture systems. Monitoring the fruit farm is one of the potential applications that may help improving fruit farm profitability through observing the fluctuation of the oriental fruit fly population and environmental conditions in the field. These data can be used to provide knowledge or warning for the farmers and government officials to accurately respond to the variations in the field. In this study, a remote agro-ecological monitoring system is presented to be a context-aware sensor platform to analyse the relations between population dynamics of the flies in the field. The system consists of three major layers, i.e., the front-end sensing layer, the telecommunication layer, and the data collection and analysis layer. A Global System of Mobile Communication (GSM) module is used to enhance the ubiquitous monitoring capability of the system. The monitoring system has been deployed to investigate the population dynamics of *B. dorsalis* since August 2008. Historical sensing data is available through a web-based decision support program built upon a database and a pest population forecast model, so that farmers and government officials are able to receive real-time farm status, as well as to carry out pest control program. Compared

*Journal of Communication, Navigation, Sensing and Services, Vol. 1,* 67–92.
doi: 10.13052/jconasense2246-2120.114

with the previously version of the system, various useful functions have been added into the proposed system, and its accuracy has been improved when measuring different parameters in the field. We believe that the proposed system provides a valuable framework for farmers and pest control officials to analyse the relation between population dynamics of the fruit fly and meteorological events. Based on the analysis, a better insect pest risk assessment and decision supporting system can be made as an aid to IPM programs against *B. dorsalis*.

**Keywords:** ICT, remote monitoring, pest control, agro-ecological monitoring, wireless sensor networks.

## 1 Introduction

The oriental fruit fly, *Bactrocera dorsalis* (Hendel), is one of the most economically important phytophagous insects around the globe [1–3]. There are nearly 5000 documented species of the fruit flies (family Tephritidae) distributed around the world, of which 148 species pose potential threats to the agricultural systems in Taiwan [4]. A female oriental fruit fly may lay about 1200 to 1500 eggs during her lifetime. It takes 14 to 24 days for the fruit fly to develop from egg to adult and it lives for up to 3 months. Such a growth rate allows the oriental fruit fly to complete about 8 to 9 generations per year. More than 150 species of fruit and vegetables, including those economic important ones, such as guava, peach, mango, and citrus, have been recorded as the host plants of the oriental fruit fly. The fruit is rotten and dropped if getting attacked, so the quality and quantity of infested fruit are declined. Thus, in Taiwan, the economic turmoil caused by the oriental fruit fly reaches up to 130 million U.S. dollars per year [5–9].

The oriental fruit fly was firstly recorded in 1912, and it mainly originated from Taiwan and Ryukyu Islands [10]. Due to quarantine failure, the fruit fly has spread to most of the countries or regions around the Asia-Pacific areas because of increasing activities in national/international trading since the past century [11–14]. Such problems further enhance the dispersal capability of the oriental fruit fly, and also result in more successive biological invasions by the pest [15]. For the protection of agricultural products, as well as for the protection of human health and the environment from misused pesticides, the government of Taiwan annually sets a budget of 5 million U.S. dollars to carry out Integrated Pest Management (IPM) programs in order to reduce the economic loss caused by the oriental fruit fly [16, 17]. The IPM is an integrated approach that consists of a series of actions, including pest identification, risk assessment, threat prevention, and agro-ecological problem solving, without using a large quantity of chemical pesticide when agricultural systems are deployed. Attempts have also been made to investigate more favourable approaches to control the population of the oriental fruit fly, such as physical, biological [18, 19], integrated [19] and chemical [20] controls, sterile-insect [21, 22], male annihilation [23, 24], RNAi [25], and bait application [26, 27] techniques. Some studies also suggest that early resistance management programs should be initiated in order to restore the efficiency of pesticides and to reduce the growth of resistant strains [28, 29]. In some control programs, an additional investment in labour to count and record pest numbers may be worthy in surveying major pests in high value crop sites. As in Taiwan, a large number of monitoring stations have been set up [30, 31]. However, such a surveillance approach relies on manual measurement at a 10-day scale, and the data is often incomplete and inaccurate without dense temporal resolution and ambient data. Using such data to determine whether the oriental fruit flies are normally present or in an outbreak status may lead to a wrong conclusion. Nor can the data be used to analyse the factors that involved population dynamics of the oriental fruit fly, because the life cycle of the oriental fruit fly might be shorter than the survey interval in the summer. Furthermore, past studies have demonstrated that the ecology of the oriental fruit fly is influenced by the factors like temperature, solar illumination, rainfall, and different kinds of crops [32–38]. In order to have greater knowledge regarding the connection between these factors and the population dynamics of the oriental fruit fly, it is necessary to develop an automatic surveillance technique which can provide accurate, long-term, and up-to-the-minute data (e.g. meteorological data, population dynamics of pests, etc.) when monitoring a fruit farm. This new precision agriculture technique makes it possible to assess the risk of insect pest outbreaks in the fruit farm and arrange better pest control activities. Hence, the IPM programs can be greatly improved by the technique [39, 40].

## 1.1 Programs for Oriental Fruit Fly Control in Taiwan

Taiwan is an island with 36,000 square kilometres and a wide range of terrain. With warm climate and abundant rainfall, many types of fruit and vegetables grow profusely on the island, and several are indigenous to other countries. Unfortunately, up to 89 types of fruit and vegetables in Taiwan are the potential hosts of the oriental fruit fly. Female oriental fruit flies deposit their eggs in the pericarps, and then the larvae feed and grow inside the fruit. Pest infestation not only causes severe economic damage by reducing both the production and quality of the fruit, but also impedes exports of fresh fruit due to quarantine restrictions imposed by other countries [5–9]. In the past decades, the male annihilation with poisoned attractant (i.e., methyl eugenol) was widely adopted by the government of Taiwan for its large-area control strategies against the oriental fruit fly. However, it is very difficult to achieve satisfactory control, since the oriental fruit fly is already widely distributed throughout the island with an extremely board host range, including 29 non-economically important plants [3]. These plants are unworthy for growers but the government has to initiate costly pest control programs for them to prevent further spread of the oriental fruit fly to non-infested areas. Many protective strategies, including fruit bagging, net-house cultivation, and pesticide sprays, have been widely used by growers to protect their agricultural products against the oriental fruit fly. However, these methods are costly and require much manpower, and pesticide sprays could seriously affect the environment and human health.

Since 1994, the Taiwan Agricultural Research Institute (TARI) has published a pest information bulletin every 10 days regarding the population size and distribution of the oriental fruit fly in Taiwan, and the bulletins, serving as reference tools, are sent to fruit growers to take necessary action to control the population of the pest. The monitoring data and analysis reports in the bulletins come from more than 77 monitoring stations, each of which includes 9 spots equipped with fly traps that contain the attractant mixed with insecticides, and thus a total of 613 monitoring traps have been assembled and placed at major farm production regions where soils, climate, and environmental parameters vary [30, 31]. Such information is proven useful in identifying locations with high-density of the oriental fruit fly. However, without the aid of the automatic recording technology, the monitoring system relies on manual measurement at a 10-day scale without meteorological information. The data collection is often incomplete and requires much manpower and high management costs [7] In addition, the long sampling interval makes the monitoring system not fully capable of assessing insect pest risks. Nor does the system send out warning messages regarding instant pest breakouts.

Recently, the Bureau of Animal and Plant Health Inspection and Quarantine (BAPHIQ) of Taiwan has launched a series of research projects to overcome the shortcomings of the manual pest surveillance methods. These projects focus on assisting IPM programs against the oriental fruit fly by (1) developing a remote sensing technique to monitor pest migration in selected areas; (2) avoiding unnecessary pesticide spraying to reduce the harm brought to the environment and human health; and (3) improving the competitiveness and the economic diversification of locally grown fruits (Chang et al., 2010). Since 2000, the BAPHIQ and TARI have successfully organized 155 farmer associations to collaboratively participate in IPM programs to control the population of the oriental fruit fly. Furthermore, modern technologies developed in recent years have also been incorporated into pest control programs [19]. For example, the geographic information system (GIS) was used to visualize the acquired data and to assess the damage caused by the oriental fruit fly [41, 42]. In addition, a new type of protein bait, GF-120 [23, 43, 44], has been introduced to Taiwan since 2003 for female fruit fly density control from spring to summer [45]. Spraying GF-120 on the crops inside and outside orchards decreased the fruit damage ratio from 70% in 2005 to 15% in 2006, and the GF-120 spray also reduced the environmental hazard caused by chemical pesticides. These successful results have drawn much attention from foreign research groups that would like to learn from the Taiwan experience in pest control [46].

An IPM system is generally designed with three major components: inspection, identification, and control. In precision agriculture, the most important task of the program is to perform pest inspection and identification. A successful IPM program relies on frequently visual inspection and accurate pest identification. When the population of the pest reaches an unacceptable level, mechanical control methods are the first options to disrupt the breeding of the pest. The surveillance and control methods described above are still insufficient to guarantee a successful IPM program for two reasons. Firstly, monitoring the population of the oriental fruit fly at a 10-day scale provides very limited information to model the population dynamics of the pest. Secondly, applying biological bait and chemical pesticides should be considered as a last resort for pest control, since the method might affect local environments.

## 1.2 Historical Background on WSN-based Monitoring Systems in Taiwan

To yield solutions to the problems mentioned above, the National Science Council (NSC), TARI and BAPHIQ have cooperated with our research team at National Taiwan University (NTU) to develop a remote pest monitoring system targeted at the oriental fruit fly since 2006, 2009 and 2011, respectively. In 2008, the authors presented a prototype of the system that is able to provide precision agriculture services. It is able to automatically report the environmental conditions and the number of trapped pests in real-time [7].

The acquired data was stored in a database for further analysis. The prototype system has been deployed and tested in an experimental farm at the NTU campus, and the experimental results show that the system can effectively reduce the cost of labour and increase the effectiveness of IPM programs.

## 2 Overall System Architecture

Monitoring of the population dynamics of the oriental fruit fly was conducted by deploying the proposed agro-ecological monitoring system to the crop sites of interest. To increase the applicability of the monitoring system, since 2008 the prototype system has been extended to a large-scale, long-term and real-time agro-ecological monitoring system designed to monitor various types of pests. The monitoring system has been deployed to 20 crop sites that cover different terrain in Taiwan, and the system consists of 12 WSN-based monitoring stations (currently assembled by 163 sensor nodes) and 3 standalone monitoring stations. Detailed information regarding these monitoring stations is shown in Fig. 1.

The overall configuration of the proposed remote agro-ecological monitoring system is depicted in Fig. 1. The fundamental architecture of the monitoring system can be divided into three major layers: the front-end sensing layer, the telecommunication layer, and the data collection and analysis layer. The purpose of the front-end sensing layer is to acquire field data from the area

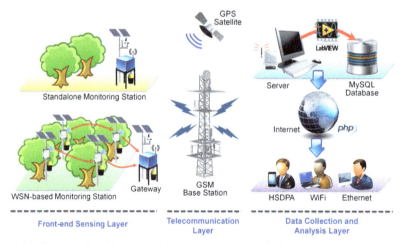

**Figure 1**   Conceptual architecture of the remote agro-ecological monitoring system presented in this study.

of interest. Two monitoring systems – a standalone monitoring station and a WSN-based monitoring station – are deployed in this study, and the detailed information regarding their configurations will be addressed in the following section. In the telecommunication layer, the sensing data measured by the sensors in the front-end sensing layer is organized into text messages for data transmission using the short message service (SMS) via GSM [47] which is virtually accessible to people from any populated areas in the world. Finally, in the data collection and analysis layer, the server built upon LabVIEW [48] receives the sensing data acquired from all monitoring stations via GSM. All historical data is stored in a MySQL [49] database for information retrieval and analysis by specialists via web services programmed by PHP [50]. The technologies used by the components in the infrastructure of the latter two layers are well-known commercialized technologies. Due to limited space, further discussion on these components is omitted, and the following discussion will mainly focus on the devices used in the front-end sensing layer, which make major contributions to agro-ecological monitoring. Designed for different field conditions and to meet farmers' demands, two types of sensing approaches are available to monitor the crop sites of interest. One is a standalone monitoring station, and the other is a WSN-based monitoring station.

## 2.1 Standalone Monitoring Station

The standalone monitoring station is an evolutionary version of the remote monitoring platform presented in our previous study [7]. Different from the preceding version, the standalone monitoring station is designed on the basis of an MSP430 microcontroller (MSP430FG4619 made by Texas Instruments, Inc.). It works alone in the monitoring area, and is equipped with a set of meteorological sensors, including temperature and humidity sensors (SHT71 with high measurement accuracies $\pm0.4°C$ for temperature and $\pm2\%$ for relative humidity, made by Sensirion, Inc.). Moreover, it is also coupled with an automatic pest counting trap designed for the oriental fruit fly, a GSM module (Fastrack Supreme 10, produced by Wavecom Co., Ltd.), a GPS receiver (GM44, made by San Jose Navigation, Inc.), as well as a solar photovoltaic panel (its power rating is 20 W) and a battery (the battery voltage is 12 V with the energy storage capacity equal to 100 Ah). The design for the automatic pest counting trap will be discussed in detail later. Fig. 2 shows the external configuration and internal architecture of a standalone monitoring station deployed in Chiayi. The standalone monitoring station aims at capturing the event of the fruit fly crawling into the trap. The number of the fruit flies and

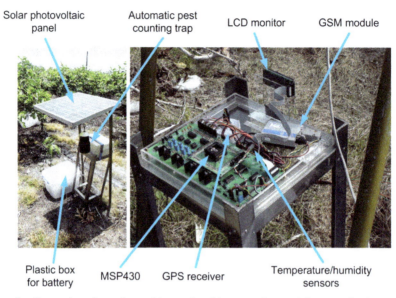

Solar photovoltaic panel

Automatic pest counting trap

LCD monitor

GSM module

Plastic box for battery

MSP430

GPS receiver

Temperature/humidity sensors

**Figure 2**   External configuration and internal architecture of a standalone monitoring station deployed in Chiayi.

real-time readings acquired from all meteorological sensors are sent to the back-end servers via GSM every 30 minutes.

## 2.2 WSN-based Monitoring Station

In addition to relying on single-point sampling using the standalone monitoring station, a WSN-based monitoring station is employed in this study to provide a unique, wireless, and easy solution to tackle distributive and multiple-point agro-ecological monitoring tasks over an area of interest with better spatial and temporal resolutions. Each WSN-based monitoring station is composed of a number of wireless sensor nodes and a gateway. The design and implementation of the wireless sensor nodes are addressed in the following section.

### 2.2.1 Wireless Sensor Node

In a WSN-based monitoring station, each wireless sensor node is made upon the foundation of a ZigBee transmission module (Octopus II, developed by NTHU). The ZigBee transmission module is coupled with an automatic pest counting trap, an 8051 microprocessor-based pest counting controller, an infrared interrupter controller, a luminance sensor, temperature and humidity sensors, as well as a solar photovoltaic panel (its power rating is 20 W) and a small package battery (battery voltage is 12 V with energy storage capacity equal to 36 Ah). Fig. 3 shows the external configuration and internal architecture of a wireless sensor node and its peripheral devices located at the campus of the National Taiwan University. The wireless sensor nodes are deployed in the crop field of interest to measure the environmental conditions and the population density of the oriental fruit fly. Average distance between any paired sensor nodes is around 20 meters due to the limitation of low-power wireless transmission. All readings are sent to the gateway of the network via an ad-hoc mechanism. In addition, all measurements, including temperature, humidity, light intensity, rainfall, wind direction and pest numbers are used to analyse the correlation between ambient factors to the pest population dynamics.

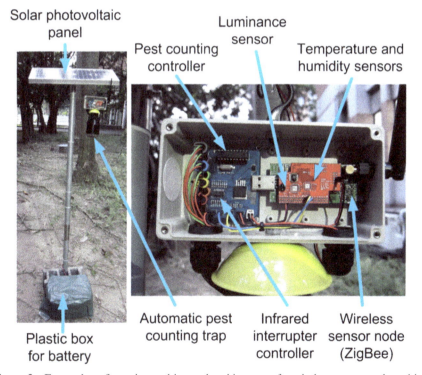

**Figure 3** External configuration and internal architecture of a wireless sensor node and its peripheral devices deployed at the campus of the National Taiwan University.

## 2.2.2 Wireless Gateway

Based on different power consumption requirements, two types of gateways are designed – MSP-based gateways and PC-based gateways. The MSP-based gateway is similar to the standalone monitoring station but without the automatic pest counting trap. The MSP-based gateway is equipped with a ZigBee transmission module (Octopus II, developed by NTHU) [51] such that it can collect all sensing data measured by the wireless sensor nodes in the agro-ecological monitoring network. Fig. 4 shows the external configuration and internal architecture of an MSP-based gateway in the network (No. 21) deployed in Pingtung. In contrast, the PC-based gateway is built upon the basis of a mini-laptop personal computer (PC), and it offers a broad range of functions designed and optimized for agro-ecological field surveillance applications. A PC-based gateway consumes more power than an MSP-based gateway, so the power for the former is supplied by commercial electricity. Each PC-based gateway is equipped with a professional weather station (WS-2308, made by La Crosse Technology, Ltd., which provides readings of an anemometer, a wind vane, a pluviometer, indoor/outdoor temperature/humidity sensors, and an atmosphere pressure meter), a ZigBee transmission module (Octopus II, developed by NTHU), as well as a GSM module (Fastrack Supreme 10, produced by Wavecom Co., Ltd.) and a GPS receiver (GM44, made by San Jose Navigation, Inc.). Fig. 5 shows the external

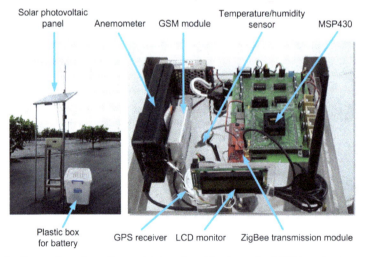

**Figure 4**  External configuration and internal architecture of a MSP-based gateway in the network (No. 21) deployed in Pingtung.

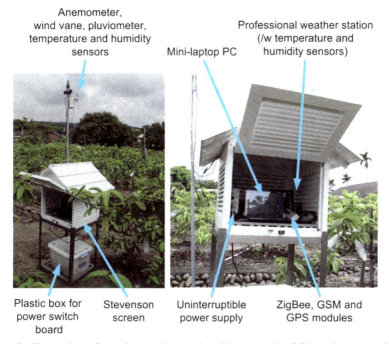

Anemometer, wind vane, pluviometer, temperature and humidity sensors

Mini-laptop PC

Professional weather station (/w temperature and humidity sensors)

Plastic box for power switch board    Stevenson screen    Uninterruptible power supply    ZigBee, GSM and GPS modules

**Figure 5** External configuration and internal architecture of a PC-based gateway in the network (No. 19) deployed in Changhua.

configuration and internal architecture of a PC-based gateway in a network (No. 19) deployed in Changhua. The purpose of a gateway is to manage wireless sensor nodes in a network, and to collect information measured by the nodes. The gateway also measures local environmental conditions using its own high precision meteorological sensors, and then periodically sends the readings acquired from the entire network to the back-end servers via GSM every 30 minutes. Detailed information regarding the sensor validation and system-level experimental results, please refer to [52].

## 3 User Survey Results

Currently, the proposed agro-ecological monitoring system has 24 registered users, and they are either the owners of the orchards or pest control officials from the experimental farms listed in Table 1 of the original paper. Among these users, 14 of them (58.3%) are senior fruit growers, and the rest (41.7%) are pest control officials with the government. A closer review of the registered user profiles revealed that 25%, 58.3% and 16.7% of the users have participated in this project for at least 2, 3, and 4 years, respectively. In addition, the user age profile showed that increasing use of the system is associated with user's age, to the point where 58.3% of the users were older than 50 years old (Fig. 6). The educational background of the registered users reveals that 45.8% of the user received bachelor degrees or higher, which is similar to the proportion of the users who serve as pest control officials (Fig. 7). Furthermore, 29.1% of the registered users are female, and the rest 71.9% are male.

In September, 2011, all registered users – including fruit growers and pest control officers – are required to complete a survey in order to evaluate the satisfaction of using the proposed system. In the proposed system, the monitoring data collected by each monitoring station was sent back to the back-end servers for storage and analysis. The analytical results of hot-spot analysis [7, 53–55] and pest density forecasting are provided by the web-based decision support program of the proposed system. First of all, the authors were wondering if the deployment of the wireless sensor nodes in the farms may cause inconvenience to the daily operations of the fruit growers in the ordinary farms (not experimental farms). However, the survey results show that 57.1% of the fruit growers stated that the deployment of the system did more good than obstructing the daily farm operations, and that the rest thought that the wireless sensor nodes did not disturb their daily operations.

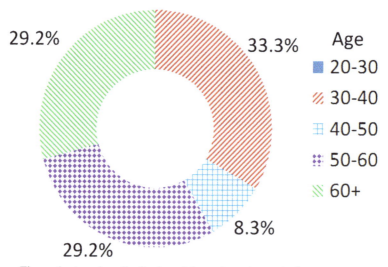

**Figure 6**   Age class distribution of the registered users in the user survey.

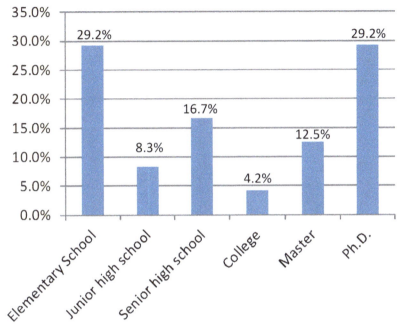

**Figure 7**   Education background of the registered users in the user survey.

Furthermore, 91.6% of the users found that the results of hotspot analysis is helpful for the farmers to identify potential breeding location of oriental fruit flies, and 95.8% of the users agreed that the pest density forecasting service provides a great help to prevent pest outbreaks (Fig. 8).

Additionally, the proposed system allows users to access to the sensing data via three different ways: short-message service (SMS), smartphone application, and computer web browser. Users' familiarity with the Internet-capable devices and interface convenience are two primary factors to attract farmers and pest control officers to make the best use of the proposed system. Among the users, 47.1% preferred to receive real-time sensing data and alert messages via SMS since more than 70% of them were familiar with ordinary cell phone, and they also thought that using ordinary cell phone was a less complex way to acquire information from the proposed system. Computer was the second preferred platform (41.2%) because about 57.1% of the users reported that they were familiar to personal computers, while 64.3% claimed that they felt comfortable to access to the web-based decision support program via personal computers. Surprisingly, only 11.8% of the users were willing to use smartphone as the medium to access to the monitoring data provided by the

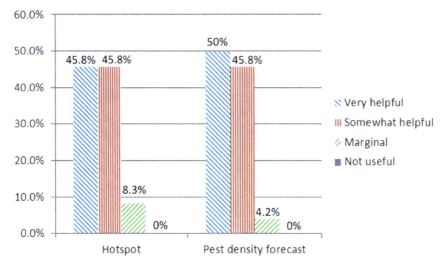

**Figure 8**   User feedback to the question of what function provided by the system changes users' management strategy.

proposed system, because there was a significant portion of the users (71.4%) who did not have a mobile smartphone yet, no wonder why using a smartphone to acquire pest-related information was not a preferred way for the users (Figs. 9–11).

With the data and information provided by the web-based decision support program, the ways through which the proposed system supports farm management practices were – avoiding unnecessary pesticide sprays (85.7%), using pesticide at the right moment (78.6%), providing a clar-ified IPM schedule (64.3%) and strategy (50%), and offering a low-cost monitoring platform (21.4%) (Fig. 12). In terms of who should pay for the proposed system after the development grants have expired, 58.3% of the users suggested that the central government should continue to sponsor the project, 25% and 16.7% indicated that local authorities and farmers' asso-ciation should support the project, and 16.7% felt that part of the funding should be supported through user fees (Fig. 13). The survey result indi-cated that the reasonable installation fee would be around 1550 USD per station if the users had to pay for the installation of a monitoring station (with 6 to 8 wireless sensor nodes), and the average user fees suggested in this survey was around 25 USD per user/month. However, all of the farmers clearly stated that the installation fee should be covered by the government.

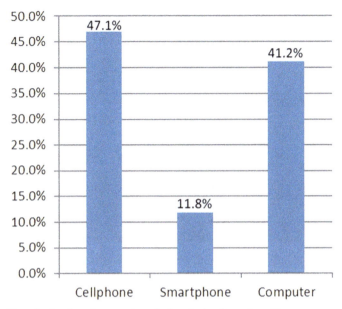

**Figure 9** User feedback to the question of what device is preferred by users to access to the sensing data.

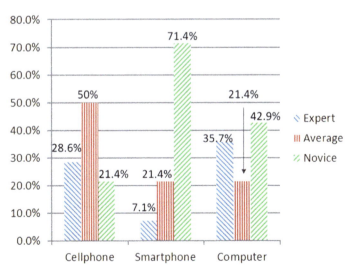

**Figure 10** User feedback to the question of how much familiarity users have with Internet-capable devices.

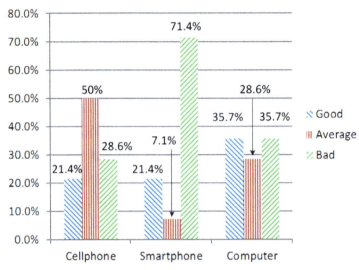

**Figure 11**   User feedback to the question of how convenient the interface provided to users is.

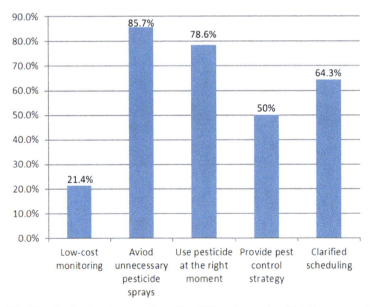

**Figure 12**   User feedback to the question of the different ways in which the proposed system have helped users manage their orchards.

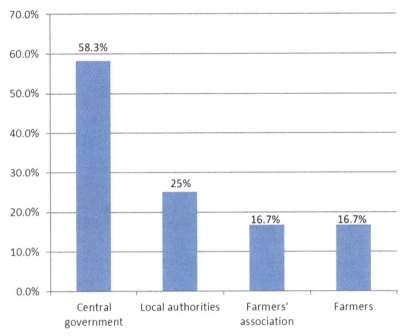

**Figure 13**   User feedback to the question of who should financially support the proposed system after the development grants have expired.

## 4 Conclusions

In this study, an agro-ecological monitoring system deployed in 20 different fruit orchards to monitor local meteorological and pest information is presented. Different from the prototype system in our previous paper [7], wireless sensor networks are integrated into the system to provide large-area monitoring capability. The proposed system contains three major layers: front-end sensing layer, telecommunication layer, and data collection and analysis layer. In the front-end sensing layer, there are two types of monitoring stations that were deployed, one is standalone monitoring station formed by an MSP-based sensing device, and the other is WSN-based monitoring station constructed by a number of wireless sensor nodes and a gateway. Both types of the stations are able to provide periodical measurement of meteorological conditions and pest information every 30 minutes. Meteorological and pest information is transmitted to the servers in the data collection and analysis layer via the telecommunication layer. All sensing data is open for public access via the Internet.

Comparing with our previous system [7], several improvements have been made to the proposed system. The previous system was a MSP-based monitoring device that is able to transmit the sensing data to a remote database via GSM platform. In this paper, a novel system is presented. Wireless sensor network technology is integrated to allow the farmers to investigate the behaviour of the oriental fruit fly using high temporal and spatial resolution data obtained by the system. The automatic pest counting trap is redesigned in order to improve the pest counting accuracy, and the evaluation results have been summarized in [55]. These improvements guarantee the practicability of the proposed system that can be deployed for long-term agro-ecological monitoring tasks in wild fields.

In order to achieve real-time system management, a remote control interface has been implemented to allow system administrators to remotely reconfigure the parameters of the proposed system via the Internet. Furthermore, a web-based decision support program is available for farmers and pest control officials to perform data inquiry, analysis, and receiving newly announced pest control tactics using any Internet-connected devices (e.g. computers, laptops, or smart phones) virtually from anywhere. Based on the feedbacks obtained from the user survey, the proposed system poses a great potential for the farmers and pest control officers to take proper precautionary actions to prevent possible pest outbreaks from getting out of control.

## 5 Acknowledgements

This work was supported in part by the National Science Council of the Executive Yuan and the Council of Agriculture of the Executive Yuan, Taiwan under contracts: NSC 98-2218-E-002-039, NSC 99-2218-E-002-015, NSC 100-2221-E-002-015, NSC 100-2221-E-027-073, 98AS-6.1.4-FD-Z1, 99AS-6.1.5-FD-Z1, and 100AS-6.1.2-BQ-B2. This work was also supported by National Science Council, National Taiwan University and Intel Corporation under Grants NSC 99-2911-I-002-201, NSC 100-2911-I-002-001, and 10R70501. The authors would like to thank the team members that include Prof. Fu-Ming Lu, Prof. Jyh-Cherng Shieh, Prof. Tung-Chung Wang , Prof. Kuo-Chi Liao, Dr. Kun-Yaw Ho, Dr. Yu-Tang Hung, Pao-Liang Chen, Prof. I-Yuan Chuang, Dr. Tzu-Rong Tsai, Wuu-Huan Shyu, Dr. Chien-Chung Chen; Ph.D. students, Yi-Jing Chu, Kelvin Jordan Liu, Yung-Cheng Wu; graduate students, Zong-Siou Wu, Kuang-Chang Lin, Chen-Ying Lin, Chu-Ping Tseng, Shieh-Hsiang Lin, Chih-Hung Hung, Jinng-Yi Wang, Chang-Wang

Liu and Tzu-Yun Lai; research assistants, Mu-Hwa Lee, for their valuable contributions to this work.

## References

[1] Vargas, R.I., Miyashita, O., Nishida, T. (1984). Life history and demographic parameters of three laboratory-reared *tephritids* (Diptera: Tephritidae). Ann. Entomol. Soc. Am., 77, 651–656.

[2] Armstrong, K.F., Carmichael, A.E., Milne, J.R., Raghu, S.R., Roderick, G.K., Yeates, D.K. (2004). Invasive phytophagous pests arising through a recent tropical evolutionary radiation: the *Bactrocera dorsalis* complex of tropical fruit flies. Annual Review of Entomology, 50, 293–319.

[3] Chen, P., Ye, H., Liu, J. (2006). Population dynamics of *Bactrocera dorsalis* (Diptera: Tephritidae) and analysis of the factors influencing the population in Ruili, Yunnan Province, China. Acta Ecologica Sinica, 26, 2801–2808.

[4] Lin, M.Y., Chen, S.K., Liu, Y.C., Yang, J.T. (2005). Pictorial key to 6 common species of the genus *Bactrocera* from Taiwan. Plant Prot. Bull., 47, 39–46.

[5] Metcalf, R.L., Metcalf, E.R. (1992). Fruit flies of the family Tephritidae. In: Metcalf RL, Metcalf ER (Eds.), Plant Kairomones in Insect Ecology and Control, Chapman & Hall, New York, USA, 109–152.

[6] TACTRI/COA, On web of plant protection manual, (in Chinese) URL: http://www.tactri.gov.tw/htdocs/ppmtable/frother-22.pdf, accessed on: Sept. 1, 2009.

[7] Jiang, J.A., Tseng, C.L., Lu, F.M., Yang, E.C., Wu, Z.S., Chen, C.P., Lin, S.H., Lin, K.C., Liao, C.S. (2008). A GSM-based remote wireless automatic monitoring system for field information: A case study for ecological monitoring of oriental fruit fly, *Bactrocera dorsalis* (Hendel). Comput. Electron. Agric., 62, 243–259.

[8] Chen, W.S., Chen, S.K., Chang, H.Y. (2002). Study on the population dynamics and control tactics of oriental fruit fly (*Bactrocera dorsalis* (Hendel)). Plant Prot. Bull., 50, 267–278.

[9] Hung, Y.T., Tsai, W.H., Kuo, K.C. (2008). Oriental fruit fly management in Taiwan: current and future. Proc. of the International Symposium on the Recent Progress of Tephritid Fruit Flies Management, 5–9.

[10] Drew, R.A.I., Hancock, D.L. (1994). The *Bactrocera dorsalis* complex of fruit flies (Diptera: Tephritidae: Dacinae) in Asia. Bulletin of Entomological Research, 2, 1–68.

[11] Fullaway, D.T. (1953). Oriental fruit fly (*Dacus dorsalis* Hendel) in Hawaii. Proceedings of the Pacific Science Congress VII, 4, 148–163.

[12] Christenson, L.C., Foot, B.H., 1960. Biology of fruit flies. Annual Review of Entomology, 5, 171–192.

[13] Hsu, E.S. (1973). Biological studies on oriental fruit fly (*Dacus dorsalis*). II. The biological effects of temperature and humidity on Oriental fruit fly (Dacus dorsalis Hendel). Plant Prot. Bull., 5, 59–86.

[14] Ye, H., Liu, J.H. (2005). Population dynamics of *Bactrocera dorsalis* (Diptera: Tephritidae) in Xishuangbanna of Southern Yunnan. Chinese Journal of Applied Ecology, 16, 1330–1334.

[15] Roderick, G.K. (2003). Tracing the origin of pests and natural enemies: genetic and statistical approaches. in: Genetics, Evolution and Biological Contro, ed. By Ehler LE, Sforza R, Mateille T, CABI Publishing, Wallingford, UK, pp. 97–112.

[16] Liu, Y.C. (2002). A review of studies and controls of oriental fruit fly (*Bactrocera dorsalis* (Hendel)) and the melon fly (*B. cucurbitae* Coquillett) in Taiwan (Diptera: Tephritidae). Proc. Symp. Insect Ecology and Fruit Fly Management, 1–40.

[17] SPC (2008). On web of plant protection service: pest advisory leaflet no.40, fruit fly control methods for pacific island countries and territories. URL: http://www.spc.int/lrd/index.php?option=com_docman&task=doc_details&gid=10&Itemid=66, accessed on: Oct. 16, 2009.

[18] Funasak, G.Y., Lai, P.Y., Nakahara, L.M., Beardsley, J.W., Ota, A.K. (1988). A review of biological control introductions in Hawaii: 1890 to 1985. Hawaiian Entomol. Soc., 28, 105–160.

[19] Orankanok, W., Chinvinijknl, S., Thanaphum, S., Sitilob, P., Enkerlin, W.R. (2007). Area-wide integrated control of Oriental fruit fly *Bactrocera dorsalis* and guava fruit fly *Bactrocera correcta* in Thailand. In: Verysen, M.J.B., Robinson, A.S., Hendrichs, J. (Ed.), Area-Wide Control of Insect Pests from Research to Implementation, Springer, Netherlands, pp. 517–526.

[20] Vargas, R.I., Ramadan, M., Hussain, T., Mochizuki, N., Bautista, R.C., Stark, J.D. (2002). Comparative demography of six fruit fly (Diptera: Tephritidae) parasitoids (Hymenoptera: Braconidae). Biological Control, 25, 30–40.

[21] Hsu, J.C., Feng, H.T. (2000). Insecticide susceptibility of oriental fruit fly (*Bactrocera dorsalis* (Hendel)) (Diptera: Tephritidae) in Taiwan. Chin. J. Entomol., 20, 109–119.

[22] Balock, J.W., Burditt, Jr. A.K., Christenson, L.D. (1963). Effects of gamma radiation on various stages of three fruit fly species. J. Eco. Entomol., 56, 43–46.

[23] Orankanok, W., Chinvinijkul, S., Sittilob, P., Thanaphum, S., Sutantawong, M., Enkerlin, W.R. (2005). Using area–wide sterile insect technique (SIT) to control two fruit fly species of economic importance in Thailand. Proc. Intl. Symp. New Frontier of Irradiated food and Non-Food Products.

[24] Wang, X.G., Messing, R.H. (2006). Feeding and attraction of non-target flies to spinosad-based fruit fly bait. Pest Management Science, 62, 933–939.

[25] Shelly, T.E., Edu, J., Pahio, E. (2007). Condition-dependent mating success in male fruit flies: ingestion of a pheromone precursor compensates for a low-quality diet. J. Insect Behavior, 20, 347–365.

[26] Chen, S.L., Lu, K.H., Dai, S.M., Li, C.H., Shieh, C.J., Chang, C. (2011). Display female-specific doublesex RNA interference in early generations of transformed oriental fruit fly, *Bactrocera dorsalis* (Hendel). Pest Management Science 67, 466–473.

[27] McQuate, G.T., Cunningham, R.T., Peck, S.L., Moore, P.H. (1999). Suppressing oriental fruit fly populations with phloxine B-protein bait sprays. Pesticide Science, 55, 574–576.

[28] Stonehouse, J., Afzal, M., Zia, Q. (2002). "Single-killing-point" field assessment of bait and lure control of fruit fly (Diptera: Tephritidae) in Pakistan. Crop Prot., 21, 651–659.

[29] Jin, T., Zeng, L., Lin, Y., Lu, Y., Liang, G. (2011). Insecticide resistance of oriental fruit fly, *Bactrocera dorsalis* (Hendel) (Diptera: Tephritidae), in mainland China. Pest Management Science, 67, 370–376.

[30] Kanga, L.H., Pree, D.J., van Lier, J.L., Walker, G.M. (2003). Management of insecticide resistance in Oriental fruit moth (*Grapholita molesta*; Lepidoptera: Tortricidae) populations from Ontario. Pest Management Science, 59, 921–927.

[31] Cheng, E.Y., Hwang, Y.B., Kao, C.H., Chaing, M.Y. (2002). An area-wide control program for oriental fruit fly in Taiwan. Proc. Symp. Insect Ecology and Fruit Fly Management, 57–71.

[32] Chu, Y.I., Chen, C.N., Chih, C.J., Wu, W.J., Yang, E.C., Hsu, J.C., Cheng, Y., Chen, C.J., Kao, J.H. (2010). Evaluation of sterile insect techniques to control oriental fruit fly in Taiwan. Research Report 98AS-9.2.1-BQ-B2(6), Bureau of Animal and Plant Health Inspection and Quarantine, Taiwan.

[33] Agricultural Research Institute, Ten-day bulletin of essential insect pests of vegetables and fruits. Council of Agriculture, Executive Yuan, Taiwan. (in Chinese) (2010) URL: http://www.tari.gov.tw/taric/modules /icontent/ index.php?op=explore&currentDir=51, accessed on Oct. 7, 2010.

[34] Arai, T. (1976). Effects of temperature and light-dark cycles on the diel Rhythm of Emergence in oriental fruit fly, *Dacus dorsalis* Hendel (Diptera: Trypetidae). Japanese J. Appl. Entomology and Zoology, 20, 69–76.

[35] Chen, C.C., Dong, Y.J., Li, C.T., Liu, K.Y., Cheng, L.L. (2006). Movement of oriental fruit fly, *Bactrocera dorsalis* (Hendel) (Diptera: Tephritidae), in a guava orchard with special reference to its population changes. Formosan Entomol., 26, 143–159.

[36] Phillips, T.W., Sanxter, S.S., Armstrong, J.W., Moy, J.H. (1997). Quarantine treatments for Hawaiian fruit flies: recent studies with irradiation, heat and cold. Proc. Annu. Intl. Res. Conf. Methyl Bromide Alternatives and Emissions Reductions, 117–1–117–2.

[37] Dobzhansky, T., Pavan, C. (1950). Local and seasonal variations in relative frequencies of species of Drosophila in Brazil. Journal of Animal Ecology, 19, 1–14.

[38] Wolda, H. (1978). Fluctuations in abundance of tropical insects. American Naturalist, 112, 1017–1045.

[39] Valadão, H., Hay, J., Tidon, R. (2010). Temporal dynamics and resource availability for drosophilid fruit flies (Insecta, Diptera) in a gallery forest in the Brazilian Savanna. International Journal of Ecology, 2010, 152437.

[40] Begon, M., Townsend, C.R., Harper, J.L. (2006). Ecology: From Individuals to Ecosystems (4th ed), Blackwell: Oxford, UK.

[41] Jones, V.P., Brunner, J.F., Grove, G.G., Petit, B., Tangren, G.V., Jones, W.E. (2010). A web-based decision support system to enhance IPM programs in Washington tree fruit. Pest Management Science, 66, 587–595.

[42] Deutsch, V.C., Joumel, A.G. (1992). GSLIB: Geostatistical Software and User's Guide. Oxford Press, N.Y. Mau, R.F.L., Jang, E.B., Vargas, R.I., Chan, C., Chou, M.Y., Sugano, J.S. (2003). Implementation of a geographic information system with integrated control tactics for area-wide fruit fly management. Proceedings of the Workshop on Plant Protection Management for Sustainable Development: Technology and New Dimension, 23–33.

[43] Peck, S.L., McQuate, G.T. (2000). Field tests of environmentally friendly malathion replacements to suppress wild Mediterranean fruit fly (Diptera: Tephritidae) populations. J. Econ. Entomol., 93, 280–289.

[44] Vargas, R.I., Peck, S.L., McQuate, G.T., Jackson, C.G., Stark, J.D., Armstrong, J.W. (2001). Potential for area wide integrated management of Mediterranean fruit fly (Diptera: Tephritidae) with a braconid parasitoid and a novel bait spray. J. Econ. Entomol., 94, 817–825.

[45] Cheng, E.Y., Kao, C.H., Chiang, M.Y., Hwang, Y.B. (2005). Area-wide control of oriental fruit fly and melon fly in Taiwan. Proceedings of Symposium on Taiwan-America Agricultural Cooperative Projects, 147–154.

[46] Su, W.Y., Chen, C.N., Cheng, E.Y., Hwang, Y.B. (2003). The geographical Taiwan-America Agricultural Cooperative Projects distribution and statistical forecasting of oriental fruit flies in Taiwan. Proceedings of the Workshop on Plant Protection Management for Sustainable Development: Technology and New Dimension, 67–110.

[47] ETSI (1999). GSM 07.07. Technical Specifications. European Telecommunications Standards Institute, Sophia-Antipolis, France.

[48] Travis, J., Kring, J., 2006. LabVIEW for Everyone: Graphical Programming Made Easy and Fun (3rd ed), Prentice Hall. Part of the National Instruments Virtual Instrumentation Series.

[49] Kofler, M. (2001). MySQL, Apress, Berkeley, CA, USA.

[50] Brown, M.C. (2002). XML Processing with Perl, Python, and PHP. Sybex, San Francisco, USA.

[51] Sheu, J.P., Chang, C.J., Sun, C.Y., Hu, W.K. (2008). WSNTB: A testbed for heterogeneous wireless sensor networks. Proceedings of the 1st IEEE International Conference on Ubi-Media Computing, 338–343.

[52] Jiang, J.A., Lin, T.S., Yang, E.C., Tseng, C.L., Chen, C.P., Yen, C.W., Zheng, X.Y., Liu, C.Y., Liu, R.H., Chen, Y.F., Chang, W.Y., Chuang, C.L. (2013). Application of a web-based remote agro-ecological monitoring system for observing spatial distribution and dynamics of Bactrocera dorsalis in fruit orchards. Precision Agriculture, 14, 323–342.

[53] Jiang, J.A., Chen, C.P., Lin, T.S., Lai, T.Y., Hung, C.H., Yang, E.C., Tseng, C.L., Lu, F.M., Liao, K.C., Shieh, J.C. (2009). Pest Hot Spot Detection for the Oriental Fruit Fly (*Bactorcera dorsalis* (Hendel)) via Wireless Sensor Networks. Proceedings of the fifth workshop on Wireless Ad Hoc and Sensor Networks.

[54] Okuyama, T., Yang, E.C., Chen, C.P., Lin, T.S., Chuang, C.L., Jiang, J.A. (2011). Using automated monitoring systems to uncover pest population dynamics in agricultural fields. Agricultural Systems, 104, 666–670.

[55] Jiang, J.A., Lin, T.S., Yang, E.C., Tseng, C.L., Chen, C.P., Yen, C.W., Zheng, X.Y., Liu, C.Y., Liu, R.H., Chen, Y.F., Chang, W.Y., Chuang, C.L. (2012). Application of a web-based remote agro-ecological monitoring system for observing spatial distribution and dynamics of *Bactrocera dorsalis* in fruit orchards," Accepted by Precision Agriculture.

## Biographies

**Joe-Air Jiang** received his M.S. and doctorate degrees from Department of Electrical Engineering at National Taiwan University in 1990 and 1999. Currently, he is a professor of Department of Bio-Industrial Mechatronics Engineering and the director of Education and Research Center for Bio-Industrial Automation, at National Taiwan University. He is also a principal investigator at the Intel-NTU Connected Context Computing Center, and an IEEE senior member. His research topics focus on bio-electromagnetics, wireless sensor networks, solar generation systems, plant factory,automation in agriculture, remote sensing and precision agriculture, fault detection/classification/location, and power quality event analysis in power transmission systems. Prof. Jiang is an active researcher. He received research awards at various occasions, including the Best Paper Awards at IEEE/PES Transmission and Distribution Conference and Exhibition in 2002, Journal of Formosan Entomology in 2007, International Seminar on Agricultural Structure and Agricultural Engineering in 2007, Workshop on Consumer Electronics in 2008, Taiwan Society of Naval Architects and Marine Engineers in 2010, and Journal of Agriculture Machinery in 2012 and 2013. Prof. Jiang also received the Academic Achievement Award from Chinese Institute of Agricultural Machinery in 2010. He published over 300 papers in different journals and conference proceedings, was granted over 30 intellectual patents from U.S.A. and R.O.C., wrote five book chapters, and edited one book with Springer-Verlag. He also received the Excellence in Teaching Awards from National Taiwan University in 2002, 2012, and 2013, and an Excellent Mentor Award from NTU in 2011. Recently, he has been the principal investigator in several large-scale integrative research projects funded by the National Science Council and the Council of Agriculture of the Executive Yuan, Taiwan. He and his research team got interviewed by the BBC and Discovery channel, and his research achievements have been broadcasted around the world via BBC and Discovery Channel in 2013.

**Cheng-Long (Richard) Chuang** is a Technology Strategist at Intel Labs, Intel Corporation, U.S.A. He is also a Scientist-in-Resident of Intel-NTU Connected Context Computing Center, National Taiwan University, Taiwan. His research interests include machine-to-machine, wireless communications, computational intelligence, FPGA/SoC rapid prototyping, smart sensing and services. His research works in smart agriculture, smart grid and magnetically guided capsule endoscopy have been covered by the media, including BBC News, Discovery Channel, New Scientist and many local media in Taiwan. He received two B.S. degrees in electrical engineering andcomputer science and information engineering from Tamkang University, Taipei, Taiwan, in2003, the M.S. degree in electrical engineering from Tamkang University, Taipei, Taiwan, in 2005, and two Ph.D. degrees in biomedical engineering and bio-industrial mechatronicsengineering at National Taiwan University, Taipei, Taiwan, in 2010.

# Towards an Unified Virtual Mobile Wireless Architecture

Oleg Asenov[1] and Vladimir Poulkov[2]

[1]The Faculty of Mathematics and Informatics, St. Kiril and St. Metodius University of Veliko Turnovo, 5003 Veliko Turnovo
[2]The Faculty of Telecommunications, Technical University of Sofia, Bulgaria, vkp@tu-sofia.bg

Received September 2013; Accepted November 2013
Publication January 2014

## Abstract

In this paper the idea for a new unified wireless network architecture is presented. The need for such a new architecture is justified and its major characteristics are outlined. Natural development of Long Term Evolution (LTE) beyond 4G, network convergence and/or evolution could naturally lead the way towards such a new unified wireless architecture, but it is expected that many of the problems of the current co-existing architectures will co-exist with such scenarios. The proposed idea of the development of a new Unified Virtual Cell Architecture (UVCA) could have the potential to overcome the problems of the mobile wireless networks of today and through the development of a good migration roadmap could be a faster and successful approach.

**Keywords:** Network Convergence, Network Architectures, Long Term Evolution, Wireless Access.

## 1 Introduction

The main questions raised in this paper are related to the evolution of wireless networks and the need of a new unified wireless network architectures and infrastructures. A general overview at the current situation could lead to some simple conclusions. There are many wireless standards such as UMTS, LTE, WiMax, WiFi, Bluetooth, RFID, ZigBee, Packed Radio solutions, etc. In many

*Journal of Communication, Navigation, Sensing and Services, Vol. 1, 93–104.*
doi: 10.13052/jconasense2246-2120.115

cases these standards are related to one and the same or overlapping services and applications. Practically this could be considered on the one hand as parallel service development or parallel investment in one and the same application services, in one and the same functionalities and in the different standards. On the other hand the existence of different wireless standards leads to the necessity one device to have installed hardware for several different wireless interfaces.

Obvious thing is also the "saturation", i.e. the availability of parallel infrastructures and dense deployments of access points in one and the same locations delivering one and the same services. In some locations infrastructure density is higher than device density. Besides cost effectiveness this raises many other problems such as throughput, interference and electro-magnetic compatibility problems. Even with the application of new highly effective modulation, channel coding and interference suppression techniques in many places the existing architectures have already reached the throughput limit (Shannon limit).

Another conclusion that could be drawn is related to the strong dependence of users on service providers. Practically a service is available only if the user pays "tax" for the infrastructure, which means that if an access to two services is required, which are offered via two different independent infrastructures, the user has to pay two "taxes infrastructure".

These simple conclusions raise the questions if there is a *need for a change* in the current telecom architectures and infrastructures and if there are *driving forces* that are strong enough to push forward to such a change. For most of the developments in telecommunications the major drivers are related to the manufacturers of new devices, the developers of new applications and services, customer demands, business competition, costs, complexity and regulations.

Manufacturers of new devices and the developers of new applications constantly release newer and smarter user devices, new applications and application-driven platforms and services. These novel communicating devices and machines introduced in the network generate more and more demands and new requirements on the infrastructure. The internet enables the everyday generation of new applications and services, without any fundamental constraints to delivering them anywhere in the world to any user using any device, thus requiring from the telecommunication networks higher and higher data transmission speeds.

During the last years mobile applications and mobile social networking have surpassed pure voice services. There are problems with the ability to process the volume of data transmitted by end users, especially in larger urban areas. Broadband mobile users are expected to reach 3.4 billion by 2014, 60b in 2020. The result is an overall increase of the multimedia traffic in the wireless networks. As a consequence, Traffic Engineering and Quality of Service (QoS) issues are becoming more and more complex. Multimedia services require a new way of traffic engineering, differentiated treatment of data flows based on traffic type, QoS, customer Service Level Agreements (SLA), user activity and user preferences. Implementation of scenarios for dynamic resource management, advanced carrier aggregation, power control, etc., with competing system operators is constrained because of complexity. This all is putting enormous pressure for changes in the design of the wireless networks architecture and infrastructure.

The overall pricing of services and cost model of the networks infrastructure is also under consideration. Today users are not concerned about technological details. Their main concern is to obtain cheaper and cheaper services that are not constrained by time and location (i.e. anytime and anywhere). To satisfy customer demands network operators strive to differentiate themselves from competitors by offering unified or better functioning products. In addition to this network operators tend to become pure Service Providers (SP). The dominant SPs move towards a data-only network, carrying VoIP. They are offering voice-only services for "free" as part of bundled data packages. The requirements towards secure, reliable and efficient data transmission, data management and QoS are growing as this is a prerequisite for their domination on the market. This of course requires also good and efficient Operation, Administration, Maintenance and Provisioning (OAM&P), which brings up the necessity of a more reliable and easier for maintenance infrastructure. Mobile operators are outsourcing OAM&P and focus on becoming only service providers.

Pressure for changes in the overall model of the infrastructure and architecture of the different types of networks are also lower costs, higher throughput, lower power consumption and green issues such as electromagnetic compatibility and health. Example is the tendency backhaul networks to serve both for fixed and mobile access based on fiber optic technology and Ethernet. With mobile wireless delivering higher and higher throughput, the size of the cells decreases (femto-, pico), thus changing and spreading the fiber network infrastructure to small cell locations. Regulation and standardization also could act as driving forces for changes in the infrastructure, such as the proposed European shared-facilities model of "RAN (radio access network) sharing".

Having in mind all the above mentioned, the main question that is raised in this paper is if the change of the network architectures of today will be a natural process of evolution or is it justified to think towards the development of a new unified wireless architecture. The rest of the paper is organized as follows: in the next section two basic evolutionary approaches towards a new architecture are discussed, section III presents the idea of the development of a unified virtual mobile architecture and some conclusions are outlined in section IV.

## 2 Evolution towards a New Wireless Network Architecture

To meet all future challenges, the expectations from a new mobile wireless architecture are generally related to the development of a big range of the access points – from 5m up to 50–100 km. ensuring extreme data rates and high traffic capacity. This supposes full integration of the fixed and mobile infrastructures and full utilization of the benefits of the different types of wireless networks which will bring to overall coverage, security, improved services and reduced costs. From the user perspective such a full integration of infrastructures means that the user will be able to connect to the network anywhere and anytime not depending on the underlying access network and will effectively utilize the infrastructure due to the implementation of a shared, flexible and reliable solution. The user will benefit also from the integration of billing and services. Billing (integrated or through bundles) must become only service dependent and must not have a network component associated with it. The introduction and integration of new services must not depend on types of networks or network architectures. Once a service is deployed over the network, it will become a service that can be run anywhere and anytime over any type of network.

From technological and service provider perspective expectation from a new architecture will give the possibility of modular and flexible territorial spread of the access, based on self-organization, self-configuration and self-regulation. This should be a unified type of dynamic spectrum access ensuring efficient and scalable utilization of the frequency spectrum. A new architecture will be cognitive and intelligent adapting to user requirements, new personalized multimedia services and applications, required QoS, etc. and at the same time "green" and cost effective, i.e. with minimized energy consumption and very-low-cost deployment and maintenance.

One possible way leading to such a new wireless architecture is the *natural evolution of LTE beyond 4G*. It is expected the unique air interface of LTE, that offers high downlink and uplink speeds, to meet the challenges of the services and the QoS requirements in the future. LTE will continue to evolve and will bring to an overall shift towards such a new network architecture. Cell sites will become smaller thus increasing capacity and lowering power consumption. LTE will become the global and unique technology for future mobile broadband as its underlying architecture is based on IP - fully packet-switched, flexible, with support of different cell sizes ranging from tens of meters, spanning femtocells and picocells to macrocells with up to a 100-km cell radius. LTE will handle the expected massive increase of users and multimedia traffic. Some arguments are given that LTE evolution will create an new LTE ecosystem – "one that is rich and open, and fosters the development of innovative mobile services and applications through the availability of new application programming interfaces which allow developers to have access to specific network assets as well as enablers such as location, presence, and security" [1].

At the first glance simple reasoning is that there should be no reason to deviate from the LTE track. Questions that could arise are related to what extend all the new applications, services and scenarios will be supported sufficiently well by the LTE evolution. Major consideration here is the issue of complexity. Very small cells in the LTE architecture means that the architecture and equipment must support complex wireless Ethernet topologies under the requirements for reliable delivery of real-time services with the necessary QoS, imposing new approaches for overall management, control, reliability and efficiency. There will be other complexity issues such as precise network timing and synchronization. For small-cell IP/Ethernet based networks that do not transport the clock reference transparently providing the necessary QoS, while maintaining delay and jitter under permissible limits to ensure recovery of the clock reference, will be quite a technological challenge [2].

Another way that could lead to a new architecture is *the fixed-mobile convergence* and the introduction of *hybrid architectures for wireless access* where the general target is the deployment of a global all-IP wireless/mobile network. Fixed-mobile convergence is practically related to the unification of wireless and wireline voice, video, and broadband data services through a seamless integration of wireless and fixed networks. Such scenarios are already being implemented as they bring advantages not only for end users, but also for service providers and operators. From a user point of view, such a convergence will bring unified billing, ubiquitous and seamless connectivity, and access to a consolidated set of services. From an operator point of view there are technical and economic benefits. They are coming from the possibility of a more effective utilization of the resources of the last mile access network and efficient use of the available public and personal, wireless local and metropolitan access infrastructures. One way towards fixed-mobile convergence is the use of fixed broadband access to backhaul mobile traffic from traditional (macro) cell sites or Long Term Evolution (LTE) evolved Node Base (eNB) stations. Another development nowadays is the introduction of smaller cell sites such as the femto cells to extend the coverage and capacity of a wireless network into homes, buildings, and closed areas, or to use hybrid architectures such as LTE and Wi-Fi. The major goal is to increase overall wireless network capacity by off-loading traffic from the full-size base stations and enabling much higher bit rates than when connected to a macro base station. The economic benefits are related to the cost-effective use of the existing, fixed broadband access network in combination with the high speeds and low latency afforded by LTE, which will enable service providers to offer integrated wireline and wireless broadband services, while lowering their operations costs. The growing demand for low-cost mobile broadband is driving the process of fixed-mobile convergence and the development of heterogeneous networks in which different radio access technologies will co-exist [3,4,5].

The main question that is raised when considering convergence is how much time will it take and will it bring to a radical change in the networks architecture as there are many technical and other problems to be solved. Key challenges are the air interface, the architecture and the protocol convergence of the different technologies. It is difficult to design a converged air-interface for different types of wireless networks due to the different coverage and channel conditions. Besides, the different networks have different signal processing capability, and different bandwidth allocation schemes. In a converged architecture the complexity should be evaluated to achieve an acceptable trade-off between the cost and performance gain. For a really converged network architecture, air-interface, protocol and control signaling should be tightly converged. In such a converged network, MAC and network layer protocols should be jointly optimized. In converged networks, the downlink and uplink control signaling should be designed, and some "cross-MAC" designs need to be implemented at the gateway. A different resource allocation scheme should be considered for converged networks, especially for scenarios where there is large number of wireless nodes with heavy traffic [6].

A related issue concerns convergence ideology. This is based on the principle of achieving the converged infrastructure through maximum reuse of the available technologies which practically hampers, or in the best case delays the creation and implementation of yet another technology. Other problems are related to the convergence of the infrastructure including costly OAM issues which have to be solved. As all the different types of future networks should be service-oriented, new usage scenarios have to be developed which could hardly be implemented directly into a converged or heterogeneous network architecture and infrastructure. Actually many of the problems of the current co-existing architectures will also co-exist in a convergence scenario and will not be solved for a long period of time.

## 3 Migration towards a New Unified Wireless Architecture

Considering the problems with network convergence a possible approach is the migration towards a new wireless network architecture. The gains from a new architecture are expected to be similar to those what we gained from cable internet: access anytime and anywhere; much more consolidated and cost-effective investments in the quality of the access infrastructure; decrease of prices of access and services; increase of the overall effectiveness in the industrial segment, due to the lower expenses for OAM&P of the infrastructures. This new architecture could be focused on Services, Applications and Content, will incorporate the architectures of device-to-device and machine-to-machine types of communications, and will drive the path towards Future Internet.

The general characteristics that such a new unified architecture must have are:

To have a new unique and unified infrastructure for wireless access, organized through clear rules for hierarchy, territory, frequency allocation.

To fully implement the proposed European shared-facilities model of "RAN (radio access network) sharing" with shared infrastructure, shared backhaul, and shared cells.

To be accessed via one unified interface in the terminal equipments – computers, tablets, smart phones, and why not also industrial controllers, identification tags, etc.

To be fully IP based, including mobile telephony.

To be self-planning, self-organizing, self-monitoring, self-regulating architecture in respect of power control, automatic switch-off of zones or groups of access points when no users or requests for service are available allowing for significantly reduced emphasis on manual intervention.

To be flexible, cognitive, intelligent in order to ensure dynamical development and evolution and employ all the advantages of cognitive radio networks.

To be fully cross-layer designed architecture with the possibility of "virtualizing" different networks and substructures (operators, private networks, etc.) in order to avoid the proliferation of various types of small-cell infrastructure equipment.

There are many challenging tasks that have to be solved for the practical implementation of such a process of migration towards a new unified architecture. Some of the challenges related to Research and Development are:

R&D of a cognitive, self-reconfigurable, self-optimizing, architecture depending on location, user activity and required QoS.

R&D of an Unified Virtual Cells Architecture (UVCA), build over the existing multi operators' Physical Cells (Base Stations) architectures. An interesting task will be the definition of optional secondary role-based virtualization of the UVCA by: QoS, by Privacy (Private UVCA), by availability for requesting temporary services (On-Demand UVCA). In this aspect another issue will be to ensure the transparency and continuity of services provided on secondary role-based virtualization trough a set of Unified Virtual Cells. The way to such a Unified Virtual Cells Architecture is to aggregate the coverage and services provided by the multi operators' Physical Cells based on one territory. This should not be a "mechanical" aggregation; a Physical Cells "Team Working Algorithm" (TWA) must be taken under R&D. During the first step of migration to a UVCA, the Physical Cells TWA will work in parallel with previous conventional features under user selection, such as the choice to be a unified user or operator dependent user. The second step will be setting-up the secondary role-based virtualization for unified use, but the option for operator depended use will be still available. At the third step there will be only unified use.

R&D of a Unified Virtual Base Station as a basic granule of UVCA based on the aggregation of the existing physical set of base stations and on the Physical Cells TWA – which will be the core algorithm of UVCA.

R&D of a technology for self-organization, self-monitoring and self-control, for an energy saving "green" infrastructure.

R&D of methods for virtualization of the existing technologies and services to be used for different applications (virtual applications) and also for group access to different services (virtual operators).

R&D of Media Access method enabling migration to multi-access domains and permitting optional or conditional user subscriptions to different access sub-domains.

Besides the R&D challenges there will be many other issues to be resolved such as the formulation of new technical conditions and requirements for spectrum use and allocation. New models of ownership and management of resources including, but not limited to, combinations and bundles of spectrum, equipment, processing, storage and energy resources, must be proposed. The architecture must be developed in such a way in order to ensure a clear plan for migration from the existing wireless architectures towards a Unified Virtual Mobile Wireless Architecture (UVMWA).

Possible roadmap for the practical implementation of the UVMWA is the following:

Stage 1. Establishment of a management Layer of the UVMWA, including primary & secondary UVMWA border session controllers (servers) per unified SOA area;

Stage 2. Attachment of the mobile operators' "Soft Switches" (switch environment) to the management layer;

Stage 3. Upgrade of base station and wireless access points middle ware with the SOA agent for UVMWA;

Stage 4. Upgrade the user wireless terminals OS with SOA client for UVMWA;

Stage 5. Start Up the UVMWA hybrid services;

Example of Start Up procedure: {.......? Primary Border Session Controllers On Service; ? Primary Border Session Controllers Interconnect; ? Operators VOIP Soft Switches On Service; ? Base Station & Wireless Access Points Service Path On Service; ? SOA Agents for UVMWA to Primary Border Session Controllers On service; ? User Equipment Configuration; ? Offering Hybrid Services for End Users.}

Stage 5. Parallel Operation: Traditional and UVMVA;

Stage 6. Switch to UVMWA.

## 4 Conclusion

In this paper the idea for the migration towards a unified wireless architecture is discussed. The main questions raised are related to the necessity and the driving forces for such a new architecture. The major arguments in its favor are related to the facts that the available mobile services and mobile contents are growing faster than mobile networks. The result will be that the building up of different parallel wireless data infrastructures for users to access same services and content could never be under consideration. Users, manufactures and service providers will be major drivers towards such a new architecture. Users want to be connected anytime and anywhere, while manufacturers of mobile devices, operators and service providers want to make the cost savings. Network convergence and/or evolution could lead the way towards such a new unified wireless architecture, but the (proposed in this paper) idea of the development of an Unified Virtual Cell Architecture has the potential for a faster migration. The development of a good migration roadmap with several stages could be a successful approach to new generation mobile networks beyond 4G, with novel approaches to resource management, traffic engineering and meeting the future challenges of multimedia services and QoS requirements.

## References

[1] Harstead E., Menendez H. The-Evolution-and-Promise-of-LTE. OSP Magazine, 2012. http://www.ospmag.com/issue/article/The Evolution and Promise-of-LTE

[2] Mudoi U. Understanding Small-Cell Unification's Vital Role In LTE And 4G. August 02, 2012.

[3] Fixed and Mobile Networks: Substitution, Complementarity and Convergence, OECD Digital Economy Papers, No. 206, OECD Publishing (2012). http://dx.doi. org/10.1787/5k91d4jwzg7b-en.

[4] Cellular and Wi-Fi: A Match Made in Heaven? Signals Ahead. March 2012, Vol. 8 No. 4.

[5] Raj M., Narayan A., Datta S., and Sajal K. Das. Fixed Mobile Convergence: Challenges and Solutions. IEEE Communications Magazine December 2010; pp. 26–34.

[6] Zhang J., Shan L., Hu H, and Yang Yang. Mobile Cellular Networks and Wireless Sensor Networks: Toward Convergence. IEEE Communications Magazine March 2012; pp.164–169.

## Biographies

**Professor Vladimir Poulkov PhD**, has received his MSc and PhD degrees at the Technical University of Sofia. He has more than 30 years of teaching and research experience in the field of telecommunications. The major fields of scientific interest are in the fields of Information Transmission Theory, Modulation and Coding. His has expertize related to interference suppression, power control and resource management for next generation telecommunications networks. Currently he is Dean of the Faculty of Telecommunications at the Technical University of Sofia. Senior Member of IEEE.

**Associated Professor Oleg Asenov**, has received his MSc degree at the Technical University of Gabrovo and PhD degree at the Technical University of Sofia. He has more than 20 years of teaching and research experience in the field of Telecommunications. The major fields of scientific interest are modeling, simulation and design of computer networks based on graph theory and applied heuristics algorithms. Currently he is Associated Professor at the St.Cyril and St.Methodius University of Veliko Tyrnovo, Member of IEEE.

*Online Manuscript Submission*

The link for submission is: www.riverpublishers.com/journal

Authors and reviewers can easily set up an account and log in to submit or review papers.

Submission formats for manuscripts: LaTeX, Word, WordPerfect, RTF, TXT.
Submission formats for figures: EPS, TIFF, GIF, JPEG, PPT and Postscript.

*LaTeX*

For submission in LaTeX, River Publishers has developed a River stylefile, which can be downloaded from http://riverpublishers.com/river publishers/authors.php

*Guidelines for Manuscripts*

Please use the Authors' Guidelines for the preparation of manuscripts, which can be downloaded from http://riverpublishers.com/river publishers/authors.php

In case of difficulties while submitting or other inquiries, please get in touch with us by clicking CONTACT on the journal's site or sending an e-mail to: info@riverpublishers.com

www.ingramcontent.com/pod-product-compliance
Lightning Source LLC
LaVergne TN
LVHW012332060326
832902LV00011B/1848